MW01258615

THE ORIGINS OF AEGIS

THE ORIGINS OF AEGIS

ELI T. REICH, WAYNE MEYER, AND THE CREATION OF A REVOLUTIONARY NAVAL WEAPONS SYSTEM

THOMAS WILDENBERG

NAVAL INSTITUTE PRESS
Annapolis, MD

Naval Institute Press
291 Wood Road
Annapolis, MD 21402

Library of Congress Cataloging-in-Publication Data
Names: Wildenberg, Thomas, 1947– author.
Title: The origins of AEGIS : Eli T. Reich, Wayne Meyer, and the creation of a revolutionary naval weapons system / Thomas Wildenberg.
Other titles: Eli T. Reich, Wayne Meyer, and the creation of a revolutionary naval weapons system
Description: Annapolis, MD : Naval Institute Press, [2024] | Includes bibliographical references and index.
Identifiers: LCCN 2023048124 (print) | LCCN 2023048125 (ebook) | ISBN 9781682479230 (hardcover) | ISBN 9781682479247 (ebook)
Subjects: LCSH: Reich, Eli T. (Eli Thomas), 1913–1999. | Meyer, Wayne, 1926–2009. | United States. Navy—Guided missile personnel—Biography. | United States. Navy—Weapons systems—Technological innovations. | AEGIS (Weapons system)—History. | Ballistic missile defenses—United States. | Admirals—United States—Biography. | United States. Navy—Officers—Biography. | BISAC: HISTORY / Military / Weapons | HISTORY / Modern / 20th Century / Cold War
Classification: LCC V62 .W55 2024 (print) | LCC V62 (ebook) | DDC 623.4/51940922—dc23/eng/20240206
LC record available at https://lccn.loc.gov/2023048124
LC ebook record available at https://lccn.loc.gov/2023048125

♾ Print editions meet the requirements of ANSI/NISO z39.48–1992 (Permanence of Paper). Printed in the United States of America.

32 31 30 29 28 27 26 25 24 9 8 7 6 5 4 3 2 1
First printing

CONTENTS

PART III. ADVANCED WEAPONS SYSTEMS

PART IV. AEGIS AND AFTER

FIGURES

TABLES

PREFACE

PETER WESTRICK, in his acclaimed history of aircraft stealth, wrote about the importance of the engineer and mid-level officers who champion new military technologies. One noted historian called this "history from the middle." This aspect of military history is often overlooked vis-à-vis the more popular top-down and bottom-up approaches to the subject. The latter attempts to explain the experiences or perspectives of ordinary people, as opposed to elites or leaders; the former emphasizes elites and leaders, as opposed to average people. The primary objective of this book is to apply history from the middle to show how two mid-level officers in the U.S. Navy, Eli T. Reich and Wayne E. Meyer, influenced the successful development and implementation of technologies that culminated in the various air-to-air guided missile systems fielded by the U.S. Navy in the second half of twentieth century. This work will also serve to illustrate the different paths that could be taken to achieve flag rank in the U.S. Navy during this period. As in my previous work on Adm. Joseph Mason Reeves, I will also demonstrate how the Navy's policy of alternating duty between sea and shore commands during this period, along with the Navy's emphasis on education and training, produced well-rounded officers with exceptional leadership qualities.[1]

While researching the history of surface-to-air guided missiles and the Aegis Weapon System, I discovered the important contributions made by Reich and Meyer. Both men can be described as product champions who provided the leadership necessary to perfect the technologies and organizations that were needed to ensure that the new guided missile systems being developed for the Navy by civilian industry could be successfully deployed in the fleet.

A well-rounded biography of Rear Adm. Wayne Meyer, the driving force behind Aegis and often called "the Father of Aegis," would certainly be in order. Unfortunately, he left no papers, letters, or memoirs, and his oral history leaves much to be desired. This makes obtaining enough material for a book dedicated to his career problematic.[2]

Though Vice Adm. Eli T. Reich's impact on the U.S. Navy's surface-to-air missile systems is less known, he too is an important figure in organizing the

Navy's resources and in correcting the Navy's problems with the 3Ts while laying the groundwork for Aegis. And while he, too, left no papers, his highly informative two-volume oral history provides a wealth of information on both his leadership qualities and the details surrounding the Navy's infamous "Get Well" program. Combining the information on the careers of both these individuals provides sufficient material for a work on their mutual rise to flag rank and their efforts to perfect the technologies needed to successfully implement the new weapon systems designed to protect the fleet from air attack.[3]

While both Vice Admiral Reich and Rear Admiral Meyer had extensive careers in the Navy, the two men traveled quite different paths on the road to flag rank. Reich, a U.S. Naval Academy graduate, was a highly decorated World War II submariner who received three Navy Crosses for his actions in command of the *Sealion* (SS 315).* He fell into ordnance work when, as the executive officer of the fleet submarine *Lapon* (SS 260), he was involved in evaluating the newly developed Mark 18 electric torpedo. This experience may have been the reason he was later selected to straighten out the research department within the Torpedo Branch of the Bureau of Ordnance in 1954. Reich also commanded the *Canberra* (CAG 2), one of Navy's first guided missile cruisers. His concerns for the problems experienced by the Navy with its first-generation surface-to-air guided missiles led to his appointment as assistant chief of the Bureau of Naval Weapons for Surface Missile Systems that evolved into the Surface Missile System Project. His outstanding leadership and administrative abilities demonstrated while at the helm of this program earned him the Distinguished Service Medal. "As a result of his outstanding personal leadership," the citation states, "his profound understanding of the mechanisms of business and government, his technical perception, and an extraordinary expenditure of personal efforts, the operational readiness of the surface-launched, anti-air weapons systems of the fleet was brought to a satisfactory level on a time scale that would have not been otherwise approachable." Admiral Reich ended his career as deputy assistant secretary of defense for production engineering and material acquisition.

Wayne Meyer began his carrier in a completely different manner. He enlisted in World War II under the V-12 Program, but due to bureaucratic delays

* Unless otherwise noted, all ships and submarines named in this work were commissioned warships in the U.S. Navy.

in the program, he did not receive his commission as an ensign in the Naval Reserve until February 1946. After receiving his commission, Meyer was sent to the Massachusetts Institute of Technology (MIT) to obtain an undergraduate degree that was not included in the V-12 Program. Meyer studied electrical engineering at MIT and received a degree from that institution in February 1947. Shortly thereafter, he was transferred to the regular Navy. After receiving his commission in the regular Navy, Meyer served three tours at sea: two as an electronics officer and one as an antisubmarine warfare/combat information center (ASW/CIC) officer. He was then sent for study at the Antiaircraft and Guided Missile School at Fort Bliss, Texas, followed by similar duty at the Fleet Training Center in Norfolk, Virginia. After completing the curriculum in Virginia, Meyer was transferred to the General Line School in Monterey, California, for duty as an instructor in nuclear weapons. Meyer's second degree from MIT and a tour as the fire control and weapons officer on board the *Galveston* (CLG 3) led to his selection in July 1963 to serve as the Terrier fire control systems manager in Reich's Special Navy Task Force for Surface Missile Systems. Meyer spent the next twenty years in various shore-based managerial positions associated with the Navy's surface missile programs, raising in rank and status to become project manager for the Aegis Shipbuilding Project. Unlike Reich, who captained four combatants and an oiler in his thirty-eight-year naval career, Rear Admiral Meyer, who was one of the most technically educated officers in the Navy, never had a seagoing command.[4]

Reading those portions of the text that relates to the development of shipborne guided missiles provides a compelling narrative of the problems faced by the U.S. Navy as it deployed its newly developed surface-to-air guided missiles. It graphically illustrates how both these outstanding officers played important roles in solving the organizational and technological problems that led to the successful development and deployment of Aegis—unquestionably the most important surface weapon system deployed by the U.S. Navy during the Cold War.

As the reader goes through the text, he or she should bear in mind the following leadership skills that one authoritative study considered critical for flag officer performance:

- exercising responsibility, good judgment, authority, and accountability
- motivating, inspiring, and mentoring military personnel

- exercising good judgment, perception, adaptiveness, and common sense to integrate priorities and eliminate irrelevant information
- guiding expectations, managing risk, and achieving results
- resolving conflict and confrontation with and among superiors, peers, and subordinates in a peacetime environment
- influencing and negotiating with people at all levels.[5]

I would add a thirst for knowledge and a desire for continual learning.

ACRONYMS AND ABBREVIATIONS

AAW	anti-air warfare
AMFAR	Advanced Multifunction Array Radar
APL	Applied Physics Laboratory
ASCA	Antisubmarine Center for Analysis
ASMS	Advanced Surface Missile System
ASW	antisubmarine warfare
BI&S	Board of Inspection and Survey
BOQ	Bachelor Officer Quarters
BT	expendable bathythermograph
BuOrd	Bureau of Ordnance
BuPers	Bureau of Naval Personnel
BuShips	Bureau of Ships
BuWeps	Bureau of Naval Weapons
CC	Construction Corps
CEC	Civil Engineering Corps
CIC	combat information center
CinCLant	Commander-in-Chief, Atlantic Command
CinCLantFlt	Commander-in-Chief, Atlantic Fleet
CNO	Chief of Naval Operations
CO	commanding officer
ComCruLant	Commander, Cruiser Force, Atlantic Fleet
ComDesLant	Commander, Destroyer Force, Atlantic Fleet
COMOPTEVFOR	Commander Operational Test and Evaluation Force
ComSubLant	Commander, Submarine Force, Atlantic Fleet
ComSubPac	Commander, Submarine Force, Pacific Fleet

CSEDS	Combat System Engineering Development Site
CV	aircraft carrier
CW	continuous wave
DCP	development concept paper
DesDiv	Destroyer Division
DesLant	Destroyer Force, Atlantic Fleet
DesRon	Destroyer Squadron
DLG	guided missile destroyer
DLGN	guided missile destroyer (nuclear propulsion)
DSARC	Defense Systems Acquisition Review Council
DSOT	daily system operating test
ECM	electronic countermeasures
EDM	engineering design model
EDO	engineering duty officer
FMSAEG	Fleet Missile System Analysis and Evaluation Group
GMLS	guided missile launching system
INSUR	inspection and survey (reports)
JICPOA	Joint Intelligence Center, Pacific Ocean Area
LOFAR	low frequency analyzer recorder
NAS	naval air station
NAVAIR	Naval Air Command
NAVORD	Naval Ordnance Systems Command (replaced Bureau of Weapons, May 1, 1966)
NAVSEA	Naval Sea Systems Command (created July 1, 1974, from the merger of Naval Ordnance [Systems] Command and Naval Sea Systems Command)
NELESCO	New London Ship and Engine Company
NOLC	Naval Ordnance Laboratory, Corona
NSIA	National Security Industrial Association
NSMSES	Naval Ship Missile Systems Engineering Station

NTDS	Naval Tactical Data System
OOD	officer of the deck
OpTevFor	Operational Test and Evaluation Force
OpNav	Office of the Chief of Naval Operations
ORE	Operational and Readiness Evaluation
ORTS	Operation Readiness and Test System
PCO	prospective commanding officer
PPI	plan position indicator (on an SJ-1 radar display)
RDT&E	research, development, testing, and evaluation
SAM	Surface-to-Air Missile
SAM-D	Surface-to-Air Missile Development Project
SAR	search and rescue
SMS	Surface Missile System
SSSC	surface/subsface surveillance control
SWU	special weapons unit
TBS	talk between ships
3Ts	Terrier, Talos, Tartar
TORPAC	Torpedo Advisory Committee
TSOR	Temporary Specific Operational Requirement
WDS	weapons director system
WesPac	Western Pacific Fleet

INTRODUCTION

IN THE LAST TEN MONTHS OF WORLD WAR II, a new form of aerial attack introduced by the Imperial Japanese Navy—the kamikaze (suicide bomber)—wreaked havoc on the U.S. Pacific Fleet, sinking 45 ships and damaging 249. The kamikaze attacks foreshadowed the introduction of the anti-ship guided missile, a weapon that did not rely on a human pilot. The psychological value of anti-aircraft fire, which in the past had driven away a large percentage of potential attackers, was of limited effectiveness against the kamikaze. Unless the pilot was killed outright, his aircraft blasted out of the sky, or his controls shot away, most kamikaze pilots would be able to hit their intended targets. The effectiveness of the Japanese suicide attacks made it imperative to come up with a new weapon to counter the kamikaze.[1]

The Bureau of Aeronautics had already begun a study project for just such a weapon in the fall of 1944. It called for the design of a high-performance, liquid-fuel rocket that would follow a radar beam to its target. The project was given a high priority in February 1945. A month later, the Naval Air Material Center in Philadelphia was directed to submit a proposal for the design, development, and manufacture of fifteen prototypes of a missile subsequently named Lark (later designated SAM-N-2). Lark was not developed in time for use in World War II, and it never entered operational service. Instead, it was used extensively from 1946 to 1951 as a test and training missile. During this time frame it provided valuable experience to U.S. military personnel in the handling and deployment of surface-to-air missiles. Lark was also the first U.S. surface-to-air missile to intercept a moving target.[2]

A project to develop a surface-to-air guided missile, similar to the one started by the Bureau of Aeronautics, was also initiated by the Bureau of Ordnance. On September 1, 1944, Vice Adm. George F. Hussey, chief of the Bureau of Ordnance, requested that Dr. Merle A. Tuve* and the scientists and

* Under Tuve's leadership, the Applied Physics Laboratory (APL) of Johns Hopkins University successfully developed the proximity fuze, had just completed the Mark 57 Gun Fire Control System for 40-mm anti-aircraft guns, and was looking for more projects to work on.

engineers at the Applied Physics Laboratory (APL) of the Johns Hopkins University investigate the practicality of developing a jet-propelled guided missile capable of protecting a task force against guided missiles launched from enemy mother planes beyond the range of existing anti-aircraft weapons. Admiral Hussey's request was made in response to ideas that had been generated by APL during a study conducted by the laboratory earlier that summer.[3]

On January 11, 1945, the Bureau of Ordnance officially assigned Task F to the APL. The stated objective of the task was to develop a ramjet-powered, ship-launched, anti-aircraft guided missile based on a detailed analysis of the problem as formulated by Prof. Jesse Beams of the University of Virginia. The missile, weighing approximately 2,000 pounds with a 600-pound warhead, was to be launched by a 2,000-pound booster rocket and was expected to achieve a maximum speed of 1,250 mph. The missile would be fired to intercept a target at a 30,000-foot altitude and at a range of 20,000 yards (approximately 10 nautical miles).[4]

Administrators at Johns Hopkins had hesitations about nearly every aspect of the project, from its classified character to the unabashedly "applied nature of the research." Nevertheless, the university found the $750,000 ($12.5 million today) contract difficult to refuse. "The scope of the program," as Matthew Montoya notes in his history of the Standard Missile, "was vast. Never before had such a weapon system been developed. There was no technological base for designing a missile with the necessary characteristics: long-range guided flight at supersonic speeds. This ambitious goal required that several different technologies be explored and that a sufficient body of new knowledge be acquired to form a rational basis for engineering design."[5]

As they had done for the proximity fuze project, the Bureau of Ordnance issued additional contracts to various universities and private contractors to investigate specific assigned areas and to accumulate data on the numerous technical problems that had to be resolved in order to produce a viable weapon. Supersonic flight and the performance of a ramjet engine had yet to be developed, and many questions had to be answered before a working missile could be designed. The Bureau's goal for the first year was to determine whether a ramjet engine could generate enough thrust to overcome supersonic drag, whether supersonic control surfaces could function without generating too much drag, and what guidance at supersonic speeds entailed.

The overall plan called for the University of Virginia to investigate servo control; Princeton University, in collaboration with the Radio Corporation of America (RCA), was to explore techniques for telemetering, tracking, and radar guidance; the University of New Mexico was to study the airframe via the use of shockwave photographs produced in wind tunnels; the Farnsworth Radio and Television Company was to investigate electronic problems; and the Esso Laboratories were to take over the study of ramjet fuels and combustion. The people assigned to each of these tasks worked at APL, and the activities were coordinated by APL.[6]

Within a month, APL had dubbed the project "Bumblebee," inspired by a wall hanging in the Office of Scientific Research and Development that read as follows:

The Bumblebee Cannot Fly

According to recognized aerotechnical tests,
the bumblebee cannot fly because of the
shape and weight of his body in relation
to the total wing area.
BUT, the bumblebee doesn't know this,
so he goes ahead and flies anyway.

Tuve thought this was an apt description of the work assigned in Task F, so "Bumblebee" seemed appropriate as the project's code name.[7]

Project Bumblebee initiated the development of three shipboard surface-to-air guided missiles known as the 3Ts (Terrier, Talos, Tartar) that were widely deployed in the fleet during the early 1960s. It was during this time that Eli Reich and Wayne Meyer became enmeshed in the Navy's guided missile program.

PART I

ELI REICH

CHAPTER 1

FROM NAVAL CADET TO SUBMARINER

ELI T. REICH WAS BORN ON MARCH 20, 1915, in Astoria, Long Island, New York, to William F. Reich and Nora M. (Faffe) Reich. His father was a member of the New York City Police Department, and he grew up in a "big old house" in Elmont, New York.[1]

When Reich was a senior in high school, he decided that he wanted to attend the United States Military Academy at West Point, New York. With his father's help, he learned that while no appointment for West Point was available, there was an alternate appointment available for the Naval Academy. Like all potential candidates for appointment to the Naval Academy, Reich had to take and pass an entrance exam. Normally this would have covered four or five subjects, but the Navy approved his high school credits, so he was required only to take the substantiating exam that covered math and English. Reich took the exam in March but did not think he had done well in the math component. This proved to be the case, for when the exam results were published, he found that he had not passed. His father encouraged him to try again and arranged for Reich to attend the Dwight School for Boys in Manhattan following his graduation from high school.

The school was a private school that prepared boys for Princeton and other Ivy League schools as well as for West Point and Annapolis. Reich spent the next school year studying math and English for the competitive examinations needed to be considered for appointment to both schools. Reich took the exam for West Point and received an alternate appointment based on his test results. Reich also passed the Naval Academy entrance exam and somehow ended up with an appointment to the Naval Academy, which he entered on June 17, 1931.[2]

The year spent studying at the Dwight School proved exceedingly fruitful for Reich as he did well in both math and English throughout his four years at Annapolis. He also did well in electrical engineering. He ranked in the top fifth of his class in all three of these subjects. Surprisingly, he achieved the poorest standing in Aptitude for Service. "I won't say that I was anti-establishment, but I really was too busy to worry about the so-called 'grease marks,'" he explained. He could put up with the system and did not worry about military bearing. Reich, according to his biographical entry in the 1935 *Lucky Bag* (the Naval Academy yearbook) had little trouble with academics and had considerable time to devote to sports. He tried to make the football team in his plebe year, but threw out his shoulder, ending any chance of playing that sport. Instead, he went out for soccer and became a member of the varsity team. He also played lacrosse and water polo, and he was on the B-squad in wrestling. Midshipmen at the time were entitled to engage in various sporting activities between the end of the academic day, around 3:30, until supper formation at 6:00, as long as they were doing well academically, as this was the only time available for extra instruction. There were also pickup games on Sunday afternoons. By the evening meal on Sunday, as Reich recalled, he "was pooped out." As a third-classman, Reich also qualified as an expert rifleman. He was ranked 120 out of a class of 442 midshipman when he graduated on June 6, 1935.[3]

Reich and the other members of his class did not receive commissions upon graduation; they were instead retained at the Naval Academy for four to six weeks. This was the time, according to the "medicos," that the relaxation and freedom from the books would restore what they believed to have been a temporary setback to the student's vision. They remained at Bancroft Hall, took their meals at the regular times, and were encouraged to engage in sports. Reich played a lot of golf during this time.[4]

At the end of June or the first week in July, the entire class went through a series of eye exams that placed the graduates into three categories: the first category was for those whose eyes had completely recovered—in this category they received a regular commission and were immediately given a set of orders to a ship; the second category were for those whose eyes had not completely recovered but for whom there was hope that they would—this group was provisionally commissioned; the third category was for those whose eyes had not recovered and who were not expected to recover—the men in this category were offered commissions in the Supply Corps. Reich fell into the first category and received an unconditional commission and a set of orders to report to the heavy cruiser *Pensacola* (CA 24) then located in Seattle, Washington.[5]

In mid-July Reich flew from Newark, New Jersey, to Seattle, Washington, on a United Airlines DC-3. The twenty-two-hour flight was his first on an airliner. By the time Reich arrived on board *Pensacola*, the best jobs had been assigned to the other junior officers who had arrived ahead of him. Reich was assigned as the assistant division officer to the R Division, one of several duties supervised by the ship's first lieutenant. This division was made up of artificers (metalsmiths and carpenters) whose main job was concerned with damage control "and the so-forths" that included the cobbler, barber, laundrymen, and the tailor. Reich described it as a "hybrid ragtag outfit" composed of thirty to forty men. His day-to-day work as an assistant R Division officer involved inspecting the men's lockers, checking that their sea bags were full, and overseeing their timely study for advancement in rating. While his day-to-day duties were rather mundane, his battle station could be very exciting: he was the officer in charge of the starboard Mark 19 anti-aircraft director. This was a very important assignment that required Reich to become completely familiar with the operation and functions of the Mark 19, which was then the most advanced anti-aircraft fire control director in the fleet. It provided fire control to the two 5-inch 25-caliber heavy anti-aircraft guns on the ship's starboard side. In the event of an air attack, Reich would be responsible for defending that side of the ship.[6]

Despite the ragtag make-up of the R Division, Reich claimed that it had more spit and polish than any other division in the ship. He attributed this to the chief warrant officer in charge of the division—a carpenter named Jones, who ran it with an iron fist. Jones' methods were contrary to anything Reich had learned or believed in—until Jones went on a 30-day leave and turned the division over to Reich. Well, he thought, "I'm going to show this division a few things about how to properly run it." As the days went by, the situation became more onerous, and he could not understand why the division was not working the way he thought it should. Years later, Reich could still remember the disdain that Jones showed when he arrived back and looked at the condition of the division. Reich was not sure what lesson he had learned, but he knew that he "had a hell of a lot to learn" when it came to handling people.[7]

In February 1936 Ensign Reich unexpectedly received orders to proceed from Long Beach to San Diego via the fleet tug *Pinola* (AT 33) and to report there to the commanding officer of the *Waters* (DD 115), a flush-deck four-pipe destroyer of World War I vintage. When he arrived on the *Waters*, the only officer on board was the former engineering officer, a lieutenant who had just been promoted to executive officer. Reich, the most junior of the ship's four officers, was unceremoniously advised that he was now the ship's engineering officer. As Reich later recalled: Here I was on a ship [which was much different from the *Pensacola*] with a captain who was a lieutenant commander, just about to make commander, a lieutenant who had just made lieutenant, another ensign, and myself. So, when the captain sent for me and we had our little talk and I told him about my concern as to whether I could really discharge the duties of the engineer, he said to me, "Don't worry about that, I know enough engineering for both of us. You just go down there and do your job and if you have problems, you come and talk them over with me. You'll do all right." *Waters*' skipper was Paul Huschke, class of '17. He also had a degree in engineering from Columbia University and had been the machinery superintendent at the New York Navy Yard when *Pensacola* was being constructed. Reich admired him and learned a lot under him.[8]

To qualify as *Waters*' engineering officer, Reich had to take the Battle Fleet's engineer's course for destroyers. In addition to written assignments, which he initially resented, Reich had to study the "Manual of Engineering Instructions." It was a very thick loose-leaf book that was always being updated with corrections and additions supplied by the Bureau of Engineering. Reich thought it was an excellent resource, but he learned more practical engineering from the chiefs and first-class machinists in the ship's engineering department. Under their guidance, he learned to adjust a Kingsbury thrust bearing, how to make turbine clearances, how to re-brick a boiler, and how to inspect a smokestack. The latter was critical, because if the lining burned through, the inner stack could fall into the boiler. According to Reich, the engineering duty on the *Waters* "was a wonderful assignment for a young officer fresh out of the Naval Academy." From his previous minor duties as third or fourth assistant to the assistant first lieutenant on the *Pensacola*, Reich was now the chief engineer of a destroyer and in charge of forty-eight men—half of the ship's crew. "It was a very responsible job, but, by the same token, to keep that 'bucket' running you were on working duty from the time you got up until you went to bed."[9]

The *Waters* was one of four ships that made up Destroyer Division (DesDiv) 19. The other ships in the division were the *Talbot* (DD 114), the *Rathburne* (DD 113), and the *Dent* (DD 116). DesDiv 19 was designated as an experimental sound division because of the QC-1A echo ranging sonar that had been installed on each of the ships. The QC-1A was an improved version of the QA sonar developed by the Naval Research Laboratory introduced in 1933. When the division tested the QC-1A in the fall of 1936, the maximum detection range was 1,760 yards at 15 knots and 3,300 yards at 10 knots. After the initial tests were completed in the waters near San Diego, DesDiv 19 was ordered to Pearl Harbor, Hawaii, where it joined the Submarine Force, Pacific.[10]

The major mission of Destroyer Division 19, now known for administrative purposes as Destroyers, Submarine Force, Pacific, was to assist in the development of the QC-1A sonar with regard to its tactical uses, reliability, and maintainability. The water conditions in Hawaii are ideal for sonar testing, as the waters around the islands in the operating areas are more than 100 fathoms deep. DesDiv 19 spent most of its time working with the S-boats of the submarine force, although they occasionally exercised as an attack destroyer division with the Battle Fleet.[11]

While the *Waters* was assigned to the Submarine Force, Reich became familiar with submarine operations and was given the opportunity to take what he termed "rides" in some of the S-boats. The S-class submarines were 220 feet long and had a surface displacement of 906 tons (1,260 tons when submerged). Although these boats were designed during World War I, they did not enter service until the early 1920s. S-boats were the first U.S. submarines designed for open ocean, blue-water operations, and made up the bulk of the Navy's submarine force throughout the interwar years. While still attached to the *Waters*, Reich also took advantage of the escape tank tower located at Pearl Harbor to undertake the escape training for submarines; and he applied to the submarine school at New London, Connecticut. The training classes for prospective submarine officers at New London were very small at that time (just twenty-four officers) and were highly selective. His application was not accepted, which is not surprising given the short time that he had been in the Navy, his junior rank, and lack of seniority. Reich never discussed his reasons for applying to submarine school. What attracted him to submarines remains unknown, but as one article in *All Hands* notes, "Submariners preferred the less rigid adherence to regulations, greater challenges, and more responsibility."[12]

Toward the end of 1937, Reich was transferred to the destroyer *Lawrence* (DD 250) for duty as the communications and torpedo officer. He also became the division secretary because the *Lawrence* was the division flagship. After serving on board the *Lawrence* for six months, in June 1938, Carleton Jones, chief engineer of the destroyer *Gilmer* (DD 233) and a good friend of Reich's, asked to swap jobs. *Gilmer* was scheduled to sail to the East Coast, where she was to be decommissioned in the Philadelphia Navy Yard. Carleton had recently married a woman from Long Beach and did not want to go to the East Coast. Reich had not been back East since leaving Annapolis, and he readily agreed to Jones' request. Reich, who was a well-trained, experienced engineer familiar with four-pipers had no trouble taking over Carleton's duties on *Gilmer*, and he made an uneventful voyage to Philadelphia. Reich spent that summer at the Philadelphia Navy Yard decommissioning the *Gilmer*, which worked out well, since he was able to spend time with his parents, who frequently came down from New York. Reich also used that time to take the examination for lieutenant junior grade, which was conducted in the *Gilmer*'s chart room on the bridge.[13]

In August 1938, Reich received orders notifying him that upon decommissioning of the *Gilmer* he was to attend the next submarine class beginning in January 1939. In the interim Reich was temporarily assigned to the battleship *Texas* (BB 35). Reich was detached from the *Texas* just before Christmas of 1938 and immediately purchased a 1938 Pontiac for $860. It was his first new car. He drove it from Norfolk, where the *Texas* was based, to New York City, where he spent a few days with his parents before driving on to New London, Connecticut.[14]

When Reich started the submarine course at the beginning of January 1939, Cdr. James Fife Jr. was the new officer in charge. Reich considered him to be a disciplinarian and "extremely demanding as far as we students were concerned." Fife would later serve as chief of staff for Submarine Squadron 20 and would be awarded the Distinguished Service Medal for his work as chief of staff for the Submarines Asiatic Fleet while that force was operating against the Japanese in defense of the Philippine Islands and the Netherlands East Indies. Classes at the submarine school began on Monday mornings and continued until noon on Saturday; they included courses in electrical engineering and batteries, torpedoes, leadership ("which was the particular privy" of Commander Fife), and submarines, a course that covered the operations and physics of submarines.[15]

One former student of the school, according to Matthew McGraw, claimed "that the intellectual rigors of the school rivaled those of civilian universities. Instructors impressed upon their students from day one that the base at New London was 'a taut' school, and we are going to have to work like the dickens to succeed." Reich never commented on the difficulty of the curriculum except to say that he did not have much free time during the week as his evenings were occupied with studying for the quiz that would inevitably be given the next morning. This did not bother him because the daily routine during the week was pretty much a repeat of his days at the Naval Academy. But he and the other students detested the Saturday morning routine. Reich, like most of the other officers in the course, wanted to get away on the weekends and did not like having to spend Saturday mornings taking two-hour essay exams starting at 8 a.m., followed by Fife's hour-and-a-half long lecture on leadership. Needless to say, Fife was not very popular.[16]

After finishing the submarine training course in June 1939, Reich was assigned to the submarine *R-14* (SS 91). The 20-year-old, 600-ton submarine was one of the boats assigned to the base squadron and was used for training purposes. As Reich later noted, the school boats "were obsolete, if not downright antique." The *R-14*'s diesel engines, furnished by the New London Ship and Engine Company (NELESCO), had been built to an untried design due to the exigencies of wartime production; they were seriously flawed and suffered from torsional vibration. Reich, as the boat's engineering officer, was responsible for keeping them in working order and constantly worried about keeping the boat's propulsion machinery functional.[17]

CHAPTER 2

SEALION DUTY

IN SEPTEMBER 1939, Reich received orders to the commissioning and fitting out crew of the *Sealion*, one of the new so-called fleet boats under construction at the Electric Boat Company, at Groton, Connecticut. *Sealion*, launched on May 25, 1939, was the eighth submarine in the new *Sargo* class. These were the first U.S. Navy submarines that had enough speed, range, and endurance to operate as part of the Battle Fleet. As such, they became known as fleet submarines. *Sealion*, with a displacement of 1,450 tons and a length of 310 feet, was 60 percent longer and displaced two and a half times more tonnage than the *R-14*, on which Reich had previously served and with which he was familiar. *Sealion* was commissioned on November 27, 1939. In mid-December, Reich was on board for the boat's shakedown cruise.

Because *Sealion*, like all the Navy's new submarines commissioned in the mid-1930s, had experimental diesel engines that were under development, the boat's executive officer was also designated as the chief engineer. A more junior officer was assigned as assistant engineer to help stand watch. During the fitting out at the yard, *Sealion*'s executive officer was able to oversee the boat's engineering department, but once at sea, the executive officer—who also served as navigator, administrator, and training officer—had little time left to supervise the engineering spaces. So he appointed Reich, who had been serving as assistant engineer, as the boat's engineering officer. This was not a particularly pleasant assignment due to the ongoing problems with the four Hoover-Owens-Rentschler (HOR) main engines built by the General Machinery Corporation of Hamilton, Ohio. These engines were two-cycle, double-acting diesels built under license from the German company MAN. The HOR's dimensions were converted to the American system and more cylinders were added. The design, according to author John Alden, was rushed to meet Navy requirements and was never properly tested under service conditions.[1]

Reich called *Sealion*'s diesels "man killers" because "they were so dammed difficult to keep in operation." They were so hard to maintain that the men in the boats called them "whores" or, alternately, "rock crushers." Why they were called "rock crushers," remains unknown, but they had propensity to destroy the stuffing box. The HOR diesels were so unreliable that in 1942 they were removed and replaced in all the boats still in service. In any case, Reich found out about the problem with the main engines during the shakedown cruise, when the boat sailed from Groton, Connecticut, to New Orleans, Louisiana, by way of St. Petersburg, Florida. By the time the submarine reached New Orleans, the diesels needed major work. The *Sealion*'s engines were in such bad shape that Reich was forced to spend three or four days sitting on the floor in the engine compartment trying to put the engines back together. This was accomplished with help from the General Machinery Company and the folks from the Texaco oil company.[2]

In late February 1940, *Sealion* arrived at the Portsmouth Naval Shipyard, in Kittery, Maine, for a routine post-shakedown availability that was supposed to be "a 'tune up' process of repairing minor defects." Instead, the *Sealion* and her sister, the *Seadragon* (SS 194), needed a complete overhaul of their engines. *Searaven* (SS 196) and *Seawolf* (SS 197) were also undergoing the post-shakedown tune up at Portsmouth, but these boats had been outfitted with engines manufactured by the Cleveland Diesel Engine Division of General Motors (GM) corporation. To support the needed adjustments and overhaul of their engines, GM sent a senior engine representative, who established a base of operations in a local Portsmouth hotel. "He held court and dispensed hospitality to the skippers and ship's engineers," which, according to Reich, "created a certain amount of good feeling. GM also sent a dozen qualified machinist and engineers—including a retired machinist's mate—to help with the work."[3]

To support the much more complex overhaul of the HOR engines on *Sealion* and *Seadragon*, the General Machinery Company sent just two technical experts. Although the company's experts interpreted the technical manuals for *Sealion*'s crew and provided some assistance in locating spare parts, they did not perform any actual work, leaving Reich and the ship's company with a "do it yourself" project. The juxtaposition of the two suppliers and their divergent ways of assisting the Navy provided a learning experience for Reich, one he would remember years later, when he took charge of the Navy's guided missile Get Well program.

From Portsmouth, Submarine Division (SubDiv) 17, which comprised all four of the boats that had been undergoing repairs at the Portsmouth Naval

Shipyard, was ordered first to San Diego, by way of the Panama Canal, with an intermediate stop in St. Petersburg, Florida. From San Diego, after a ten-day stay, the division would sail to Pearl Harbor, Hawaii. Although there were no engine casualties on the *Sealion* during the voyage, there were so many minor problems that "it was work-work-work for the engine machinists," as they struggled to keep the four main engines and two auxiliary "donkey" engines online. The *Sealion*'s propulsion system was typical of those installed on all fleet submarines commissioned between 1939 and 1940. It was very versatile because (some or all) the electrical power generated by the four main and two donkey engines could be distributed through the electrical distribution switchboards to charge the batteries that supplied power to the main drive motors when submerged (or directly into the main motors when surfaced). Unless all six engines were down, there was always some form of motor power to drive the submarine.[4]

Several weeks after their arrival at Pearl Harbor in late September 1940, SubDiv 17 received secret orders to make their ships ready for full war duty. This meant that the crew had to spike the batteries and ensure that all the torpedoes were ready for warshots. Spiking the batteries was a process that improved their storage capacity by increasing the specific gravity of the sulfuric acid electrolyte. It was used only in times of war because it reduced the life of the batteries, adding extra cost to maintaining the boats. During the spiking procedure, the crew had to replace the existing electrolyte solution with the heavier acid. This was a dangerous, time-consuming process that ruined the shirts and pants of those involved. It was certainly not something the crew enjoyed.[5]

After a month's preparation, *Sealion* and the rest of SubDiv 17 sailed out of Pearl Harbor and headed for Manila Bay in the Philippine Islands. There they would join Submarine Squadron 20 (SubRon 20), stationed at the Cavite Navy Yard located in Canacao Bay, ten or twelve miles southwest of Manila. The voyage, as Reich recalled, was a "nightmare for the engineers." After only one day's run, they lost one engine. From then on, *Sealion* always had at least one engine down and sometimes two, causing continuous juggling of the engines to keep enough power available to the main motors so the boat could keep up with the division. As one of the three junior officers on board, Reich was required to stand bridge watches during the surface portion of the cruise. After getting off the bridge watch, he would spend hours in the engine room keeping enough of the engines running to provide enough electrical power to maintain the division's sixteen-knot cruising speed.[6]

The stuffing box was the Achilles heel of the *Sealion*'s HOR engines. It was one of the most important components of a diesel engine as it prevented lubricating oil from being drawn from the crankcase into the scavenge space, scavenge air from leaking into the crankcase, and contaminated, used cylinder oil from being carried downward. Adjusting the stuffing box had to be very precise. Each cylinder had a stuffing box containing eighteen metal rings surrounding a piston. If just one of the rings was not properly adjusted or well lubricated, the engine cylinder to which it was attached would start losing compression in what *Sealion*'s machinists called "the blow." Whenever the machinist's mates began to suspect the beginning of a blow, they would attempt to prevent it by squirting a mixture concocted of lubricants and alcohol into the stuffing box with giant syringes made out of bicycle tire pumps. According to Reich, they believed that injecting the fluid would remove the bits of metal or the foreign matter that was suspected of creating wear on the metal rings that surrounded the piston rod. Preventing a blow was critical, for once it started, the engine would have to be shut down to avoid scorching a piston head or tie rod, or ruining the lower cylinder head. Replacing a piston head or tie rod was backbreaking work

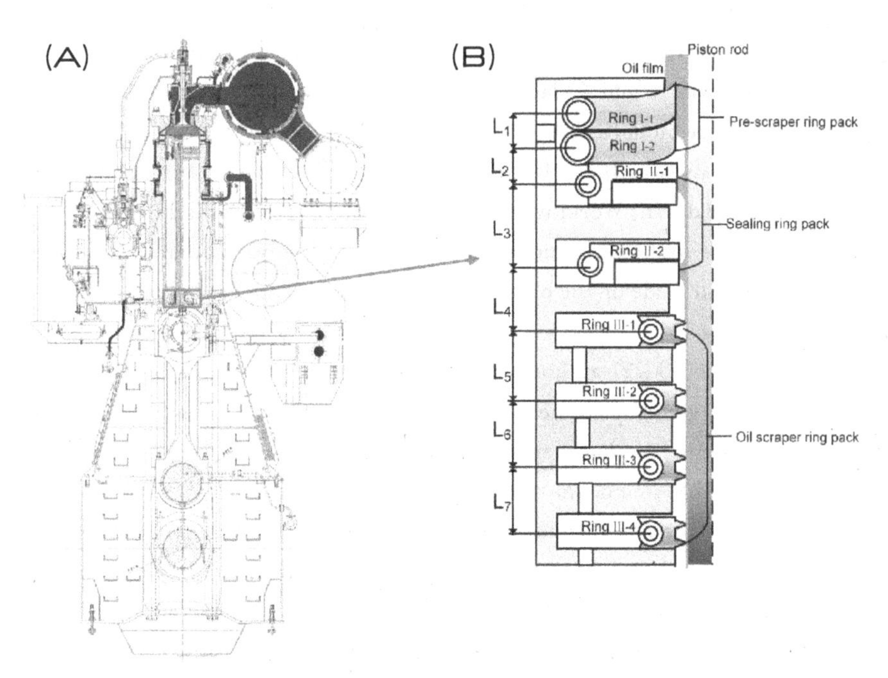

FIGURE 2.1. DIESEL ENGINE STUFFING BOX

Source: Author

that required the use of chain falls to lift the cylinder head. By the time *Sealion* entered Manila Bay in mid-November, only one main engine and the two auxiliaries were operational. "The crew," according to Reich, "was exhausted and the engines needed major repair."[7]

When the *Sealion* arrived at Cavite Navy Yard, it tied up alongside the submarine tender *Canopus* (AS 9), which was well equipped and had a good repair force. SubRon 20's engineering officer was familiar with the problems of the HOR engines and saw to it that *Sealion*'s needs received priority.[8]

Although the other HOR boats experienced problems, Reich claimed "the magnitude of the problems was always greater on *Sealion*." During one such episode *Sealion* experienced a scored main crankshaft. To fix the problem, the boat's talented machinists mates came up with an ingenious method of repairing the crankshaft in situ. They put a small grinding device on the crankshaft and used the generator as a motor to turn the shaft, deriving power from another engine or off the battery. As Reich recalled, "It was a long and tedious process, but it did the work."[9]

When *Sealion* joined SubRon 20 in Manila Bay, the squadron had more boats than any other submarine squadron in the U.S. Navy. Of the twenty-nine submarines attached to the Asiatic Fleet's submarine force stationed in Manila Bay, six were of the old S class, seven were transitional *Porpoise*-class P-boats, and twelve were modern fleet boats of the *Salmon* class.

The S-boats, which displaced approximately 1,000 tons when submerged, were designed during World War I but did not enter service until the early 1920s. They were the first U.S. submarines designed for the open ocean, blue-water operations and made up the bulk of the Navy's submarine force throughout the interwar years.

The *Porpoise* (SS 172), authorized in 1934, was the first of the so-called P-boats. With a submerged displacement of almost 2,000 tons, the P-boats were almost double the size (by displacement) of the S-boats that preceded them. The P-boats were the first U.S. submarine designed to keep up with the Battle Fleet. The Navy wanted them to be able to scout ahead of the battle line, report on the movements of the enemy, and then conduct attacks aimed at whittling down the enemy in preparation for the main engagement. The S-boats, although much more capable than any of the preceding classes, were not capable of filling this role.[10]

In 1941, when *Sealion* arrived in Manila, SubRon 20 was the largest force in the Asiatic Fleet, which, in addition to the submarines, was composed of one heavy cruiser, two light cruisers, thirteen destroyers, five minesweepers, six PT boats, and a number of auxiliaries scattered throughout Southeast Asia, from Balikpapan to Manila. When the Asiatic Fleet was established in 1902, its primary mission was of a diplomatic and constabulary nature, safeguarding American interests in East Asia. Only four of the small S-boats had been assigned to the fleet prior to 1938. Thereafter, the growing militancy of the Japanese and the potential threat to the Philippines necessitated that the Asiatic Fleet be reinforced.[11]

When the first submarines were transferred to the Asiatic Fleet in the 1920s, they were perceived as an inexpensive means of providing coastal defense that would threaten any battle force attempting to bombard the shore. Although the capabilities of the Navy's submarines had improved greatly with the *Salmon* class, they were still perceived as a relatively cheap means of defending the Philippines from a seaborne naval attack.

Sealion's mission in the spring of 1941 was to gain familiarity with the Southern Philippines, Borneo, North Borneo, and the approaches to the Dutch East Indies while also training for war. Normal operations included a six-week cruise conducted with the other submarines in her division. The division would leave Manila and head four or five hundred miles south to the island of Tawi-Tawi. The island had an excellent deep-water anchorage and was only a ten-minute run from the exercise area. "It was great for submarine operations," explained Reich. "The sea conditions were outstanding, but it was desolate," with nothing to do for recreation. To break up the monotony, the crew would leave the area on a Friday evening and spend Saturday and Sunday visiting one of the nearby towns.[12]

The primary task of the *Sealion* during these training exercises was to conduct practice torpedo firings. Torpedoes fitted with unarmed exercise heads would be fired at targets—most likely the salvage and repair ship *Pigeon* (ASR 9), or one of the Asiatic Fleet's destroyers, or perhaps one of the minelayers dispatched for this service. The torpedoes were presumably set to run deeper than the keel of the target to avoid striking the target ship. Although the warheads on the exercise torpedoes contained no explosives, their striking force was strong enough to puncture the thin-skinned destroyers or the side plating on the *Pigeon*. At the end of their run, the torpedoes, having expended all their fuel for propulsion, would become more buoyant and float on the surface. They would then be retrieved by the *Pigeon*, where they would be reconditioned and refueled to be used again.

The tactical procedure for a daytime torpedo firing, according to Reich's recollection, was a sonar approach made below periscope depth, presumably to shield the firing submarine from detection. The fire control problem was solved by using passive sonar readings to obtain target bearings until a good firing solution was obtained. At that point, a single range "ping" would be made with the active sonar to verify the range to target. Once war started, a few skippers were reluctant to use this approach since the "ping" would immediately alert any escorts of a submarine in the vicinity. Nevertheless, "This was the pro forma doctrine," according to Reich, and it had to be done. The other tactic practiced was a nighttime sonar approach in which the crew would trim the submarine until its decks were awash so that only the conning tower was visible, making it very difficult to detect. As the boat's diving officer, Reich's job was to get the boat down when ordered.[13]

In late November 1941, *Sealion* and her sister *Seadragon* (SS 194) were sent to the Cavite Navy Yard for scheduled overhaul. They had just taken everything apart when word was received that the Japanese had attacked Pearl Harbor. Reich was at the ferry landing in Manila preparing to go back to Cavite at 6:30 a.m. on the morning of December 8, 1941, when he got the news.* As Reich explained, "We were faced with a fait accompli. The war was here, and the place for a fighting submarine was not in a shipyard." Over the next two days, Reich and the rest of the *Sealion*'s crew worked day and night, making every effort to get the boat ready for sea.[14]

Reich was having lunch on the *Isabel* (SP 251) in the middle of Canacao Bay with Lt. Jack Payne, a former classmate from Annapolis, when the Japanese bombed the Cavite Navy Yard. Payne was captain of the converted yacht *Isabel,* which served as the relief flagship for the Asiatic Fleet. While Reich looked on, twenty-seven Japanese bombers flying at high altitude in three 9-plane echelons approached from the direction of Manila, passed over the navy yard to the east, and then turned back, heading directly for the yard. The first bomb, according to Rear Adm. Francis Rockwell, the commander of the 16th Naval District, exploded at 1:14 p.m. Bombs, according to Reich, "were dropping all over the place." Almost all fell within the navy yard limits, and direct hits were made on the power plant, dispensary, torpedo repair shop, supply

---

* It is impractical for the crew to live aboard a submarine while it is being overhauled, so Reich was temporarily quartered in a leased house in a Manila suburb.

office and warehouses, signal station, commissary store, receiving station, barracks, officers' quarters, and on several ships, tugs, and barges along the waterfront.[15]

Reich made it back to the *Sealion* just as the bombing ended. She was sitting in the mud and listing to port; her stern was awash over the aft torpedo room. The boat had taken two hits: one in the aft engine room and the other in the maneuvering room. Reich was extremely lucky, because had he been on board, he probably would have been killed; he normally took station in the maneuvering room while getting underway. *Sealion* was too far gone to be repaired, so orders were given to take whatever usable equipment could be removed and to destroy what was left. "The stern was deep in the water, under mud," recalled Reich, so we put two depth charges between the forward tubes and set them off."[16]

After the Japanese bombing, the Cavite Navy Yard, the largest base in the Pacific west of Pearl Harbor, was out of business. Reich, having lost his ship, was detailed as salvage officer in Cavite as assistant to Cdr. Willis E. Percifield, the division's commander who became the logistics officer for Asiatic Fleet submarines. Even before the Japanese attack, work was already underway to establish a naval base at Mariveles Bay at the southern end of the Bataan Peninsula, twenty-five miles from Cavite, on the east side of Manila Bay. Reich was ordered to move all the equipment related to submarines stored in the supply dump on Sunset Beach next to the Cavite Naval Base to Mariveles. This included all the torpedoes, mines, spare parts, and petroleum products he could find.

To accomplish this task, Reich commandeered Army trucks, cherry pickers, small tugboats, barges, and whatever else he needed to get the material from Sunset Beach, where it had been stored, to Mariveles. Reich's men worked day in and day out, Saturday and Sunday included. "We'd work until we were pooped out," Reich explained, "and then we'd get a little sleep and get up and work again." The men under Reich's command managed to move a considerable number of torpedoes and mines, along with many drums of the lubricating oil needed to keep diesel engines running. All of this was done against the background of the daily Japanese bombing raids that continued to take place in and around Manila.[17]

The following incident, relating to the bombing raids and Reich's interaction with Captain Fife, then chief of staff, Submarines, Asiatic Fleet, is illustrative of Reich's determination in getting things done and his willingness to circumvent useless rules and confront his superiors when necessary. One day, while Reich was in the process of transferring material to Mariveles, he received word that Captain Fife wanted to take a look at the logistics operation. Although Reich did not mention it, he probably was not looking forward to showing Fife around. According to Reich, "Fife wasn't the warmest person in the world . . . and when he looked at you, you always had a sense of ill feeling. Like you knew you weren't doing something right." To get around from place to place, Reich had commandeered an old Ford that he used to pick up Fife when he arrived at the Cavite ferry landing. Just as they passed the magazine inside the navy yard, the air raid sirens went off. Reich, in his oral history, tells us what happened next:

> I pulled over to the right and told Captain Fife to come on and we jumped down from the seawall, which was about six feet high, and there was this big concrete pipe (about five feet in diameter) under the road with just a little water in it. We got down in the pipe and waited for about four or five minutes, after which I popped my head out and took a look around to see what was happening. But it was obvious to me, and I knew what was going on—nothing. I knew that normally they come up with an "all clear," but I also knew that the damn "all clear" signal might not come for forty-five minutes to an hour (or even later) because those birds who were running the show down at the commandant[']s . . . were cautious to a fault.
>
> So, I said to Captain Fife, "Let's go!"
>
> He said, "What do you mean?"
>
> I said, "We have got to get the hell out of here—I want to get to Sunset Beach."
>
> He said, "Yu can't go because there's no 'all clear.'"
>
> I said, "Captain, we ain't going to do a G—damn thing in this place if we're going to pay attention to those kinds of rules. Now come on."[18]

Fife somewhat reluctantly said OK, and they continued on the inspection without any further interruptions or problems.[19]

Reich continued to work on the stores needed to keep the Asiatic Fleet's submarines in operation until just before New Year's Eve 1941, when he received word that all submarine officers not attached to boats were to be evacuated, and he received orders to board the *Stingray* (SS 186). The *Stingray*, which had just returned from her first unsuccessful war patrol, had a new skipper, Lt. Cdr. Raymond J. "Bud" Moore, a 1929 graduate of the Naval Academy who had relieved the boat's previous captain just days before.[20]

CHAPTER 3

REICH'S FIRST WAR PATROLS

ELI REICH WENT ON BOARD the *Stingray* accompanied by Lt. Frank Brown, another submarine-qualified officer marooned in Mariveles. They joined the boat as supernumeraries: they stood watches, but they had no department duties. This gave the two experienced submariners the opportunity to observe the crew during its transit to and maneuvers in the Gulf of Tonkin, where the *Stingray* was sent on patrol. After several days of observing the crew and its operation of the submarine, Reich and Brown both felt that things were not right with the crew. "People were very, very nervous and there was something very disconcerting about the general deportment—of the people on the ship—something neither one us had observed in our prior experience with subs." Of particular concern was the first-class machinist's mate who assisted the diving officer in the control room. Reich had been a diving officer for a long time and was very familiar with control room operations. He would look at the machinist's mate and just shake his head in disbelief. He never went to bed and never slept. As Reich recalls, he (the machinist's mate) was haggard looking and wouldn't leave his station in the control room where he served as the "auxiliary man."[1]

Even more disconcerting to Reich was the poor performance of the boat's torpedo fire control party. While the *Stingray* made contact with several Japanese ships and fired several torpedo salvos, they sank only one ship, the 5,120-ton passenger/cargo ship *Harbin Maru*, which went down on January 10, 1942.[2]

From the Gulf of Tonkin near Hainan Island, *Stingray* was ordered to proceed to the Makassar Straits to intercept Japanese warships headed for Balikpapan. One night, as he was getting into his bunk in the wardroom after duty on the midnight to 2 a.m. deck watch, Reich was startled when the diving alarm went off. He was thrown out of his bunk as the boat suddenly dived at a fifteen-degree angle—the steepest he had ever experienced. As he landed on the deck, a number of depth charges went off "that shook the hell out of the boat." Unbeknownst to the crew, a Japanese destroyer had been stalking the

Stingray. To escape from the destroyer, Commander Moore ordered the *Stingray* to a depth of 300 feet, which was considered deep for a boat of *Stingray*'s class. The boat stayed down for the rest of the day while the crew cleaned up the mess in the galley and the loose materials that had been thrown about in the control room. Reich observed that whatever level the tension was in the boat before this event, it was now compounded.[3]

Going down to 300 feet was a dangerous maneuver considering the material condition of the *Stingray*'s "dutchman"—the short length of piping between the outboard engine exhaust and the inboard engine exhaust valve. The dutchman was full of holes, and the only thing keeping the water out was the inboard stop valve. Paul E. Summers, the boat's engineer and diving officer, was particularly concerned about going deep. He knew better than anyone the precarious nature of their situation. The dutchman could not be fixed at sea; they would have to wait until their next port of call before it could be repaired. Summers described Reich "as a philosopher who read books all the time while on *Stingray*." Reich evidently "took pleasure in his books and was somewhat low-key and soft-spoken, but underneath the veneer was an aggressive commander."[4]

The torpedoes fired by the *Sealion* and the other submarines assigned to the Asiatic Fleet were the steam-powered Mark 10s. The Mark 10 had a 497-pound warhead packed with TNT triggered by a Mark 3 contact exploder. But its depth-setting gear was faulty, as the Bureau of Ordnance advised in a notice addressing the problem issued on January 5, 1942. When or if this information was received on board the *Stingray* remains undetermined.[5]

Reich was very familiar with torpedo doctrine, and he knew that to get hits, you had to get close. A rule of thumb regarding the accuracy of torpedo firing in those days stated if you fired at a range of 1,500 yards or less, the chances of hitting the target were two or three times better than if you fired at 3,000 to 4,000 yards of the target. And if you fired inside of a 1,000 yards, your chances of hitting the target were 50 to 100 percent better than if you fired at 1,500 yards. To get hits, according to Reich's view, you had to have the guts and tenacity to keep bearing in, despite the threat from enemy destroyers. You had to get up close if you wanted to sink ships. Thus, Reich "was disgusted" when Captain Moore fired from 4,000 yards and missed.[6]

Reich liked Moore as an individual, but he "didn't have much respect for him as a skipper." As Reich noted in his oral history, one cannot command a combat submarine in a performative manner:

> You've got a role to play, and that's a tremendous role. You not only have to have a sense of mission, but the biggest thing you have to have is an understanding of the whole environment in which you are placed—a good understanding of all the material things, the boat, the condition of things. You can't be preoccupied. You've got to know the torpedoes; you've got to know the fire control business. You absolutely have to know all those things as well as you have to know the crew and that's not easy to do sometimes.[7]

When *Stingray* reached Surabaya, Java, on February 12, 1942, Reich was sent to the Bachelor Officer Quarters (BOQ) at the Dutch submarine base to await new orders. By then Reich must have been notified of his promotion to lieutenant, effective January 1, 1942. Along with the promotion came additional responsibilities. After two days in the BOQ, he received a note from Captain Fife, who had relocated to Surabaya with the rest of the Asiatic Fleet's submarines, with new orders: he was to report to *Stingray* as prospective executive officer to relieve Hank Stunn. Reich was not very happy with the assignment. He felt like he had been stabbed in the back: Fife could have sent him "to any other bucket in the Navy," but not back to that boat. Because the Japanese were closing in, the *Stingray* was ordered to leave immediately and quickly got underway for Fremantle, Australia.[8]

When the *Stingray* arrived in Fremantle on March 3, 1942, it went alongside the submarine tender *Otis* (AS 20), where she received much-needed repairs undertaken with help from the tender's personnel and the electrical shop of the local streetcar company. *Stingray* departed Fremantle on her third war patrol on March 16, 1942; she was headed for her designated patrol area in the Celebes and Java Seas. The only worthwhile target encountered during this patrol was a Japanese destroyer cruising just off the coast of Makassar City on the southern end of Celebes Island. Although the submarine fired three torpedoes, all were misses* (one wonders what Reich thought about the setup and the performance of the sub's captain). *Stingray* returned to Fremantle on May 2, 1942.[9]

By the time *Stingray* arrived in Fremantle for the second time, she was badly in need of the kind of overhaul that could be accomplished only in a stateside yard. She got underway again on May 27, with orders to proceed on patrol to

* Presumably the crew was now aware of the March 10 depth problems and had taken measures to account for the discrepancy.

Guam by way of Davao Gulf in the Philippines, and from there on to Pearl Harbor. On the afternoon of June 28, *Stingray* was 190 miles north of Yap, halfway between Davao Gulf and Guam, when she sighted two ships with escort. The submarine closed the range to the convoy and fired four torpedoes at the first ship, sinking the converted gunboat *Saikyo Maru*. *Stingray* continued to patrol in the vicinity of Guam until June 15, when she headed for Pearl Harbor. After a short stay in Pearl Harbor, the boat was sent to the Mare Island Naval Shipyard, in Vallejo, California, for the much-needed overhaul.[10]

At the beginning of September 1942, Reich took leave and flew to New York, where he married his fiancée, Nora M. Taffe, in St. Matthews Cathedral on September 4. Reich never discussed where or how he met his wife or the correspondence that he must have had during his time at sea. In any event, Reich's honeymoon was cut short when he received orders directing him to report back to *Stingray*. As soon as *Stingray*'s overhaul was completed, Reich's next order was to report to the *Lapon* (SS 260), still under construction at the Electric Boat's yard in Groton, Connecticut, in preparation for becoming her executive officer. So, Reich flew back across the country, went out with the sea trials, and returned to the East Coast, where he picked up his new bride and headed for Groton.

Electric Boat's yard was just across the Thames River from New London, Connecticut. Although a new yard had been added and the workforce had expanded from its prewar level of four or five thousand to eighteen thousand, Reich found the same key people he had dealt with during his time with the *Sealion*. This was very useful to Reich. "If you know your way around the waterfront and if you know the government inspectors, then it's even more helpful if you get to know the company's people," he explained. *Lapon* was not very different from *Sealion*, either. She even had the same HOR engines. The submarine's crew had grown in size, however, and now consisted of eight officers and seventy-four men. Her captain, Cdr. Oliver "Ollie" G. Kirk, having previously served only in S-boats, was not familiar with the bigger boat. So, he assigned Reich the task of organizing the crew. The rest of the crew was composed of six reserve officers, or mustangs, twenty experienced seamen, and around fifty-four inexperienced greenhorns. Reich considered working the crew and commissioning the boat hard work, but he didn't encounter any problems he couldn't handle.[11]

As soon as a new boat was fit for sea duty, it would normally have been sent to the Pacific, where it was urgently needed for combat duty. But this was not the case for *Lapon*. Instead, she was sent to the U.S. Naval Torpedo Station in Newport, Rhode Island, to assist in the testing of the new electric torpedoes then under development. The big advantage of the electric torpedo was its absence of a wake, which could give away a submarine's position. Although the Navy had begun investigating the design of an electric torpedo in 1919, the project was discontinued during the Great Depression for lack of funds. It resumed in July 1941, however, when the Bureau of Ordnance set aside $500,000 to develop the Mark 2 electric torpedo. The Torpedo Station, which was working around the clock to produce badly needed steam torpedoes, lacked the resources necessary to manufacture electric torpedoes, so production was shifted to the General Electric Company. Seeking a second source, the bureau of Ordnance asked the Westinghouse Electric and Manufacturing Company for assistance. Representatives from the company met with members of the Bureau's staff on March 10, 1942. A week later, the company representatives visited the Torpedo Station to collect data and design material for the electric torpedo. After reviewing this data, which presumably included inspection of the German G7e electric torpedo that had been recovered by the British, Westinghouse officials decided that the quickest way to produce an electric torpedo was to copy the German model as closely as possible. The Bureau of Ordnance concurred and awarded the company a development contract to produce five experimental models of an electric torpedo, designated the Mark 18, for testing at the Torpedo Station by June 1942.[12]

After the short cruise from New London, where *Lapon* had undergone sea training, she arrived at Narragansett Bay and moored at the Gould Island pier. Gould Island had been acquired by the Navy in 1918 for use as a storage and test site for aircraft-launched torpedoes. In 1942 the Navy built a two-story firing pier at the north end of the island to replace the barge that had previously been used for test-firing torpedoes. Historian Brian L. Walin, explains, "The bottom floor contained four tubes for both surface and underwater torpedo launching. The second floor had an observation deck where the tests could be monitored." The complex also included a large building for assembly and testing torpedoes.[13]

Not long after the *Lapon* tied up at Gould Island, Reich, Commander Kirk, and the torpedo ratings from the boat's crew walked over to the firing complex to see what was going on there. While walking around, they discovered that a section of the building had been given over to the Westinghouse people for use in testing the Mark 18 torpedoes. "They [the Westinghouse people] were as pleased as punch to talk" with the veterans from *Lapon*'s crew, knowing that the men had a wealth of practical experience with torpedoes. In talking with Westinghouse's engineers, it became obvious to Reich that they were very anxious to get on with the testing. It was also clear that they were not getting a lot of help from torpedo experts assigned to the Torpedo Station. "What we found," Reich said later, "was simply sickening. The Mark 18 was an extremely promising weapon. But the Newport guys—steam torpedo guys—had a big NIH—Not Invented Here—attitude toward the Westinghouse torpedo." After discussing the situation with Commander Kirk, they decided, "God damn it," they "were going to make these fish run." Reich and Kirk were going to get in there and really work with the Westinghouse guys to make sure that the Mark 18 was successfully developed. They began by firing test samples while moored alongside the pier or just parallel to it.[14]

By examining the damage to recovered torpedoes that had run erratically, the developers determined that the Mark 18 torpedoes were not leaving the torpedo tubes properly. The problem—they theorized—was due to the starting torque of the motor, which caused the torpedo to "kick" the skegway that guided the torpedo along the shutter area around the outer doors as the torpedo was leaving the tube. This caused the torpedo to veer away at an angle. The proposed solution was to add a delay to the Mark 18's motor so that it would not start until the torpedo had cleared the torpedo tube. Working with Westinghouse, the team from *Lapon* helped design such a system. This solved the erratic run problem.[15]

After spending six weeks in Narragansett Bay, *Lapon* briefly returned to New London before departing for Pearl Harbor. She arrived there on June 1, 1943. By then, Reich had been promoted to lieutenant commander. Along the way Commander Kirk, with Reich's assistance, composed a damning memorandum covering the development and testing program for the Mark 18. The memo, signed by Kirk, states the following:

> Basically, it's the Bureau's fault. They sent a boy to do a man's job. An Ensign. . . . What this set-up needs is a "Pushing Officer" with lots of enthusiasm and the backing of the Bureau, and (very important) some rank. What latent enthusiasm there was here has now petered out and it won't rise again until they know their efforts are something. . . . I

> suggest for a "Pushing Officer" a submarine skipper, Commander or Lieutenant Commander, with combat experience, preferably one who has fired a few duds, and had occasion to swear at a few expeditors and bureaus as the depth charges were raining down. This officer should be familiarized with the devious workings of this queer place, then, with the authority and backing of the Bureau, he can do wonders.

Reich's experience with the Mark 18 problem at Torpedo Station and the conclusion contained in the report composed with Kirk would guide his actions years later, when he faced a similar problem with the Terrier guided missile and its cousins, Tartar and Talos. What Commander Kirk and Reich called a "Pushing Officer," would later be termed the "Product Champion" or "Product Owner."[16]

As a new boat, *Lapon* was required to undergo prepatrol training before she received orders to the war zone. This was normally a seven- or eight-day miniature patrol in Hawaiian waters supervised by a senior training officer. The training was structured to "run you ragged," recalled Reich. "Pseudo enemy aircraft would harass you and target ships would come out at uncertain times for torpedo firing exercises." The prepatrol exercises were very hectic, which Reich considered in many ways to be harder than an actual patrol. Once this training was over, *Lapon* returned to Pearl Harbor and was given a two- or three-day load-out period to fuel, take on stores, and pick up torpedoes.

CHAPTER 4

SEALION WAR PATROLS

ON JUNE 24, 1943, the *Gato*-class submarine *Lapon* (SS 260), with Eli Reich as executive officer, departed Pearl Harbor with orders to proceed to the Sea of Japan, colloquially known as the "Emperor's Bathtub." No U.S. submarine had operated in this area before. *Lapon* was one of four boats ordered by Vice Adm. Charles A. Lockwood Jr., the commander of Submarine Force, Pacific Fleet, to make this first foray into the Sea of Japan. All the boats set sail within the same twenty-four-hour period. The route took them through La Pérouse Strait that separated the Russian island of Sakhalin from the Northern part of the Japanese island of Hokkaido. Once through the strait, they were scheduled to spend four days patrolling within the Sea of Japan before exiting, again through La Pérouse Strait. Vice Admiral Lockwood considered it unsafe for them to remain any longer, fearing that the Japanese would block the strait once it was known that American submarines were operating within the Sea of Japan.[1]

After departing Pearl Harbor, *Lapon* headed northwest toward the Kuril Islands. The weather was cool as they approached La Pérouse Strait. They sailed through a dense fog, running on the surface and using the boat's SJ radar to plot a course through the strait. The SJ radar was a 10-cm surface-search set that utilized an "A" scope for range.* It was installed on U.S. submarines beginning in June 1942. Reich claimed that they could not have made landfall if they had not had the radar. "We plotted the whole eastern entrance to the La Pérouse Strait primarily on radar," he explained. The SJ radar, according to Robert Dienesch, "had the greatest impact on the submarine war," greater than any other single factor. In addition to its use as a navigation tool in foul weather, it provided a submarine's commander with a tactical picture that could be used to detect and track targets and escorts.[2]

Once in their assigned area off the Korean coast, the only ships sighted by *Lapon* were fishing boats. *Lapon* attacked one of the largest of these using her

* A plan position indicator (PPI) was later added to an improved version of this radar.

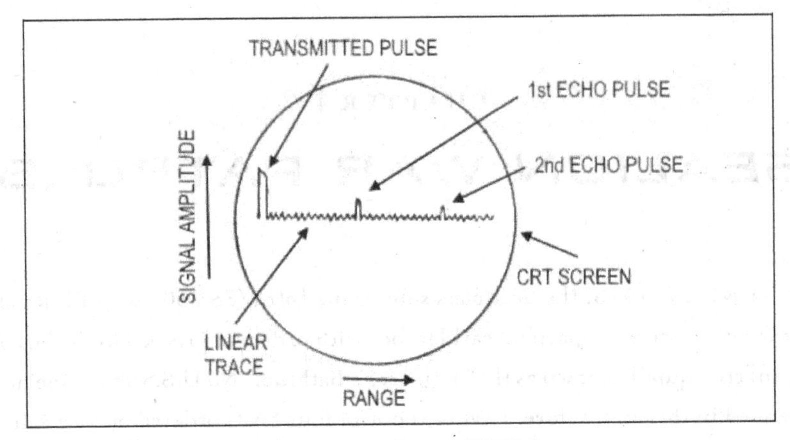

FIGURE 4.1. A SCOPE

The A Scope is the oldest and simplest radar display. It shows the amplitude of the reflected radar return on the vertical scale and the range on the horizontal scale. The bearing to the target is manually determined by physically turning the radar antenna until the maximum amplitude is reached.

Source: Author

20-mm cannon in a surface action, but little was achieved for the effort. As Reich noted, "We only scared them and revealed our presence." *Lapon* remained within the Sea of Japan for three days without seeing another ship. She then headed north and west through La Pérouse Strait and set a course for Pearl Harbor.[3]

The forays of the first four U.S. submarines into the Sea of Japan accomplished much less than Admiral Lockwood expected; but it generated consternation in Japan, which previously had assumed that ships in the Sea of Japan were safe from submarine attack.[4]

After *Lapon* returned to Pearl Harbor, Reich received a message to report to Dick Voge, his former skipper on the *Sealion*, who was now the operations officer for the Submarine Force, Pacific Fleet. Commander Voge asked Reich if he would like to take command of the second *Sealion*.

"You're kidding me," said Reich.

"No," Voge replied, "there's a new *Sealion* being built at Electric Boat, and I think you ought to be the first skipper."

That is how Reich, with a little help from Commander Voge and Lt. Cdr. Frederick B. "Freddie" Warder, a former classmate and highly successful submarine commander who worked for Voge, persuaded Admiral Lockwood to approve Reich's orders to proceed to New London, Connecticut, as prospective commanding officer of the second *Sealion*. The new *Sealion*, which was under construction, was a *Gato*-class submarine similar in design to *Lapon*. Reich was overjoyed at the news. He had been the executive officer on two boats and was chafing at the bit for his own command. In addition, he had married the previous September and was looking forward to the opportunity of rejoining his wife.[5]

Reich arrived in New London in early October and was able to supervise preparations for the boat's launch on October 31, 1943. From that point until *Sealion* (SS 315) was commissioned on March 8, 1944, the commissioning detail, under Reich's command, underwent an intense period of training. In addition to Reich, the officers assigned to the commissioning detail consisted of four seasoned officers who had experience in war patrols and three junior officers fresh out of submarine school. Twenty percent of the enlisted personnel were also veterans. The remainder were submarine-school graduates.[6]

When Reich arrived in New London, the head of the training command for new construction was Capt. Leon F. "Savvy" Huffman. Reich described him as "tough disciplinarian who had high standards" but was unfamiliar with the way things were being done in the Pacific. Training under Huffman's aegis was not a pleasant experience for Reich or for some of the other captains of the new boats constructed at the Electric Boat yard.[7]

Very early in the game, Reich and Lt. Cdr. Edward N. Blakely of the *Shark* (SS 314) had to contend with what they considered were arbitrary orders from the training command on how to formulate the boat's organization. Captain Huffman had gone over the organization put together by Lt. Cdr. Bliss C. Hull, captain of the *Perch* (SS 313), which was eight weeks ahead of the *Sealion*. The organization of a submarine was not a trivial matter. A lot of it had to do with the constitution of the fire control party and the manner in which the ship's watch were handled. Neither Reich nor Blakely liked the way Hull had laid out the organization. They did not want anything to do with it and expressed a great deal of unhappiness about it to one another. It took several discussions before they came to realize that "there was no point in fighting 'city hall' so long as we could train our crew as fast as we could." Besides, they knew that once their boats were in the Pacific, they could do what they wanted.[8]

Following a shakedown cruise off New London, *Sealion* sailed for the Pacific, arriving at Pearl Harbor on May 17, 1944. Reich and his crew spent another three weeks in precombat training before departing Pearl Harbor on June 8 and heading for the war zone. Reich experienced his first combat action in command of a submarine after spotting what appeared to be a large fishing trawler just south of Tokyo on the way to *Sealion*'s assigned patrol area in the East China Sea. Reich did not want to waste a torpedo on such a small target, so he surfaced the boat and ordered the *Sealion* to prepare for surface action. Once surfaced they discovered not one, but two of these "trawler-like" ships. As the gun crew manned the 4-inch deck gun and *Sealion* began closing on the target, it become clear that these ships were not trawlers but Japanese patrol boats similar to the U.S. Navy's submarine chasers. Although Reich believed that he had them outgunned, he wisely retreated when 40-mm shells began dropping around *Sealion*. As he said, "It was two against one, so I decided to get the hell out of there."[9]

Reich's second combat encounter did not turn out well, either. On June 23, during their second or third day of submerged patrolling in the East China Sea, they spotted a small Japanese warship with a high stack that resembled the profile of a *Sacramento*-class U.S. gunboat. After maneuvering into perfect position for a good firing setup, Reich ordered three Mark 14 torpedoes to be launched. He then watched through the periscope as all three torpedoes ran under the target without exploding. Later, Reich could not remember the depth settings he had ordered, but the Mark 14 was notorious for running too deep, despite the Bureau of Ordnance's ongoing attempts to rectify the Mark 14's problems. The gunboat immediately swung around and started running down the torpedo wakes leading to the *Sealion*. Reich ordered the boat down to 400 feet and, with the help of the newly developed Depth Charge Direction Indicator,[†] was able to outmaneuver the Japanese warship as she pounded the area with depth charges. *Sealion* "took a shellacking" for the next several hours but received no serious damage and was able to resume her patrol.[10]

On June 28, *Sealion* caught and sank the 2,386-ton naval transport *Sansei Maru* near Tsushima Island. Two days later she sank a sampan using her deck guns. In July Reich moved *Sealion* closer to China's coastline to patrol the approaches to Shanghai. In the early morning hours of July 6, *Sealion* intercepted a convoy south of Four Sisters Island and commenced firing torpedoes at two of the merchantmen in the formation. Within minutes, the 1,922-ton *Setsuzan Maru* sank,

† The Depth Charge Direction Indicator was a sonar device used to indicate the general direction of depth-charge explosions.

and the convoy scattered. *Sealion* retired to the northeast to evade the convoy's escort destroyer, which had begun to search for the submarine. Reich fired four torpedoes at the approaching destroyer but missed. Nevertheless, these actions illustrate Reich's fearlessness and the fighting spirit that his Annapolis classmates had noted in the *Lucky Bag*.[11]

Sealion sank two more freighters on July 11: the 2,417-ton *Tsukuski Maru* No. 2 and the 1036-ton *Taian Maru* No. 2. The latter involved three attacks over a seven-hour period. In the third and final attack of the engagement, Reich, realizing that he had only one torpedo left, moved as close to the target as he could get in order to ensure that he secured a hit. After such a long chase, the tenacious Reich did not want to lose his prey. He ordered the last torpedo fired when the *Sealion* was just 1,000 yards from the target. It struck the freighter amidships, causing an explosion that showered the *Sealion* with debris. Having expended all *Sealion* torpedoes, Reich was ordered to Midway, where the submarine tender *Fulton* (AS 11) would rearm it and make any repairs needed. After arriving, the crew was allowed to make use of the recreation facilities established on the island while Reich was flown to Pearl Harbor to be briefed for his next patrol.[12]

Although the results of Reich's first war patrol as a submarine commander were not as spectacular as the ten ships sunk by Lt. Cdr. Richard O. Kane during his second war patrol in March of 1944, the four ships sunk by *Sealion* were above average for the number of ships (3.5) sunk per patrol by the top ten submarine captains in World War II. Reich later received the Navy Cross for his actions during this war patrol. It a was common practice during World War II for this decoration to be awarded to the commanding officer of any vessel or aircraft that was responsible for sinking an enemy combatant.

After four and a half weeks of refitting and rest and another prepatrol exercise, *Sealion*, with Reich in command, departed Midway on August 17 to begin her second war patrol. *Sealion* headed for her assigned patrol area in the Luzon Strait, south of Formosa, in a loosely coordinated patrol with the *Growler* (SS 215) and *Pampanito* (SS 383). In the predawn hours of August 31, *Sealion* conducted a night surface attack against a Japanese convoy and damaged a tanker before coming under attack from the deck guns of other ships in the convoy. Still surfaced, Reich evaded the convoy's escorts, maneuvered ahead of the convoy, and submerged. At 0730, Reich fired three torpedoes at one of the escorts that looked like an enemy destroyer. Two of torpedoes struck the Imperial Japanese Navy minelayer *Shirataka*, which sank later that morning. Later that day,

Sealion fired three more torpedoes at another escort, which had spotted the torpedo tracks, turned to avoid them, and attacked the *Sealion* with depth charges without causing any damage to the boat. After evading the destroyer, *Sealion*, low on fuel and torpedoes, was ordered to Saipan in the Mariana Islands, which had been secured by the Marines in July. She was to be refueled and rearmed in Tanapag Harbor by the submarine tender *Holland* (AS 32) on September 7. *Sealion* arrived in the morning, loaded up, and departed that evening heading back to her patrol area.[13]

After *Sealion* had gotten underway, Reich received the daily ULTRA secret intelligence dispatch from the Intelligence Center at Pearl Harbor. The message alerted Reich and the commanding officers of *Growler* and *Pampanito* that a large Japanese convoy from Singapore was heading for their patrol area. The information had been obtained by the Joint Intelligence Center, Pacific Ocean Area (JICPOA) that had been established in September 1943 to provide intelligence support to wide ranging combat operations. Working with the operations people, JICPOA knew which submarines were operating in a particular area and would pick from various sources the materials that pertained to that area. The information in many cases was so precise that it included departure time, speed, and number of ships in a convoy. Intelligence was obtained, for the most part, from the Japanese Water Transportation code, which the Navy's cryptologists had broken. "By the summer of 1944 the intelligence center," according to one authority, "was functioning [as] a well-oiled machine, run like a large newspaper." Very few Japanese ships could sail without JICPOA's knowledge. Convoy HI 72, which was reported to Reich, was no exception.[14]

Reich estimated that *Sealion* was 1,700 miles from the probable point of contact with the convoy. To get there in time to make an attack, he ordered a surface run at maximum speed. On September 11, 1944, *Sealion* rendezvoused with *Growler* and *Pampanito*; together, they formed a line of search, seeking out the convoy indicated in the ULTRA dispatch. *Growler* was the first to spot the convoy. At 0200 she attacked the formation, and *Sealion* followed suit. *Growler's* torpedoes sent a destroyer to the bottom. *Sealion* fired two torpedoes in a surface attack that missed their targets; she was then taken under fire by two of the escorts. Reich ordered top speed and managed to keep ahead of the escorts until they broke off to rejoin the convoy. An hour and a half later, *Sealion* closed on the convoy, and at 0522 fired three torpedoes at a tanker, and then swung around to obtain a firing position on the last ship in the nearest column. At

0524, the tanker, possibly hit by torpedoes from both *Pampanito* and *Sealion*, burst into flames, illuminating the large transport that Reich had selected for his next target, the *Rakuyō Maru*. He fired two torpedoes at the large transport, both of which hit. *Sealion* was forced to dive deep after the attack, and after several attempts at reengaging at the convoy, cleared the area and headed north in an attempt to find it again.[15]

Unknown to Reich or the crew of the *Sealion*, the *Rakuyō Maru* was transporting 1,318 Australian and British prisoners of war who had been used as forced labor on the Thailand-Burma Railway. It was a 250-mile construction project upon which the POWs had been forced to work since June 1942. What happened to the POWs after the *Rakuyō Maru* was struck is graphically described by James Hornfischer in *Ship of Ghosts*:

> As the *Rakuyō Maru* took on water, her decks became a battlefield of sorts. Panicked prisoners rushed the ladders. The Japanese crewmen, holding them at bay with makeshift weapons, took all ten of the ship's lifeboats and made good their escape. The unlucky crew remaining on the sinking ship were left to contend with their prisoners. . . . They set upon the Japanese crewmen and beat many of them to death by hand. . . . The ship disappeared beneath the waves around 5:30 p.m. that afternoon.[16]

The survivors threw everything in the water that would float. "The Japanese," notes Gregory Michno, "were too busy saving themselves to be concerned about the prisoners, although there were instances of murder on each side." While the Japanese were searching for survivors, prisoners on one raft sat on a lone Japanese soldier and held his mouth shut so he wouldn't be found. When the Japanese left the area, the POWs beat him to death.[17]

That night, *Pampanito* caught up with the remnants of the convoy and torpedoed the *Kachidoki Maru* at 2240. The 10,509-ton transport, which sank in minutes, was carrying 6,000 tons of bauxite and more than 2,000 POWs. Those who were not killed by the torpedoes had to jump into the sea in the dark of night[18]

After attacking the convoy, *Growler* was detached from the wolf pack, and *Sealion* and *Pampanito* continued to patrol in the South China Sea. On September 15, *Pampanito* moved back to the area of the original attack and found men clinging to makeshift rafts. The submarine's crew, thinking they were Japanese, brought out small arms and got ready to take prisoners. As *Pampanito* approached several rafts they saw some of the men had Australian "Digger" hats. The men in the rafts were also waving and shouting as hard as they could. Getting the survivors on board was not an easy task. "They were very hard to handle," according to the account given by Lt. Cdr. Landon L. Davis Jr., *Pampanito*'s executive officer. "They were just covered with a heavy oil, all over their bodies, their hands, and we had a devil of a time trying to get them on board, they were slick, couldn't pick them up. They were quite weak, and they could not help themselves very much. . . . I remember the first one that came up—he actually kissed the man as he pulled him up on deck, he was so happy to get on there. They were quite in a state of hysteria, they had practically given up when they finally got picked up by us."[19]

As *Pampanito* began to pick up survivors, she broke radio silence and sent a message to *Sealion* to ask for help. Reich immediately ordered *Sealion*, which was about thirty miles to the northeast, to the scene and began to pick up survivors, too. After rescuing fifty of the survivors, Reich decided that he could take no more on board. Night had also fallen, which made additional searching difficult. *Pampanito* had already alerted submarine headquarters in Pearl Harbor of the situation and *Pampanito*, along with *Sealion*, were ordered to make for Saipan, the closest off-loading point that had facilities for handling the survivors. As *Sealion* passed through the Luzon Strait, Reich ordered the boat to slow down so that they could bury four of the POWs who had died enroute. Reich described how this was done: "We did it in the classic way—we put them in a blanker and tied a four-inch shell to each of them. Then we had a little prayer service and put them over the side." From Saipan, *Sealion* sailed to Pearl Harbor for rest, recreation, and refit.[20]

CHAPTER 5

REICH "BAGS" A BATTLESHIP

SEALION RETURNED TO PEARL HARBOR on September 30, 1944, and departed again for her third patrol on October 31. After a ten-day run, *Sealion* transited the Tokara Strait and entered her assigned patrol area in the East China Sea. During the first part of her patrol, Reich experienced three separate, harrowing incidents in one three-day period, each of which had the potential to sink the *Sealion*. The first occurred during a routine check of the torpedo firing circuits. Somehow things got mixed up between the torpedo officer in the conning tower and the torpedomen in the aft torpedo room. Instead of testing an empty torpedo tube, they closed the firing circuit on a fully loaded torpedo tube. The torpedo took off and exited the submarine taking the outer door with it. For safety reasons, the torpedo had to travel 250 yards before the warhead was armed, so there was little danger that it would explode. Without the outer door, however, the only thing keeping the sea from entering the boat was the inner door. Knowing that they were in shallow water, Reich decided to make the best of it and continue on with the patrol. He was more concerned with the loss of the torpedo and the fact that one of the torpedo tubes was now out of action.[1]

The second dangerous incident occurred while *Sealion* was patrolling off the mouth of the Yangtze River. As Reich was looking through the periscope, he saw an old-fashioned horned mine coming toward the boat. There was not anything he could do about it, except to watch it float by some seventy-five yards or so away. A day later, the battery in one of the Mark 18 electric torpedoes in one of the forward torpedo tubes caught fire, filling the forward torpedo room with smoke. Reich surfaced the boat, and after receiving instructions from Pearl Harbor, ejected the smoking torpedo while making 12 knots astern. This was done to avoid going over the torpedo in the event it dropped beneath the boat and exploded. Reich was also concerned that it might take a circular run. Backing away at maximum speed minimized the danger from either possibility.

Despite the loss of the outer door and the other dangers Reich experienced in the first few days of the patrol, he continued to search for enemy ships in the *Sealion*'s assigned area of operation.[2]

On November 21, 1944, *Sealion* was running on the surface, heading south in a smooth sea at twenty minutes after midnight when her radar operator detected what he believed was a contact at 44,000 yards (25 miles). Reich, asleep in his bunk in the captain's cabin, was immediately notified; he jumped out of his bunk and went up to the conning tower. Reich knew that the SJ radar could detect a convoy at 19,000 yards and a large, escorted warship at 25,000, but detecting a ship at 44,000 yards* seemed unlikely. So, he was skeptical of the contact report. His first thought was that they were receiving a radar return from the north point of Formosa. *Sealion*'s radar officer, Ens. Daniel P. Brooks, insisted it had to be a contact. Although Reich had his doubts, he ordered the watch to close the contact, went down to his cabin, and came back fifteen or twenty minutes later. As Capt. John J. Hammerer Jr. has noted, "the nature of war is chaotic filled with unknowns and doubts."[3]

Reich did not know it at the time, but *Sealion* had discovered a large force of Japanese warships under the command of Vice Admiral Takeo Kurita; it included three battleships, a light cruiser, and four escorting destroyers. The three battleships, *Yamato*, *Nagato*, and *Kongō*, cruising at 16 knots to save fuel, were heading to Japan to repair damage suffered at the Battle of Leyte Gulf. As *Sealion* closed the contact, the radar returns faded in and out until individual returns began to show up at 30,000 yards. At 0048 the PPI display on *Sealion*'s SJ-1 radar scope showed "pips" for what appeared to be two battleships and two cruisers. Unbeknownst to Reich, *Yamato* had detected *Sealion*'s radar, causing Admiral Kurita to order a course change that would take the convoy away from the submarine. The convoy continued to maintain its 16-knot speed with minimal zigzagging.[4]

Reich, realizing that the only way he could get into firing position was to surface, ordered the crew to their battle stations, surfaced the boat and instructed the engineering spaces to make full speed. At 20 knots, *Sealion*'s surface speed allowed Reich to overtake the convoy without being seen. By 0146, *Sealion* was parallel to the convoy, on its port beam. Her radar now showed three

---

* This was likely caused by the large radar cross section of the ships involved; it is a phenomenon known as surface ducting, which is caused when the radar signal bounces off the sea's surface.

heavy ships in line, accompanied by three escorts: one on either beam of the formation, and one on the starboard quarter. "The Japanese fleet," explains Japanese navy expert Anthony Tully, "was still not zigzagging, and was steaming blissfully unaware on course 057 as the submarine [*Sealion*] gradually edged out in front." At 0245, the boat was ahead of the convoy. Reich ordered their speed reduced and turned the *Sealion* into the convoy to obtain a firing position. Eleven minutes later, with the range to the target at 3,000 yards, he fired all six Mark 18 electric torpedoes from the forward torpedo tubes. "Noting the tendency of the destroyers to overlap on the radar with the BBs [battleships], Reich set the torpedo running depth of his first salvo at eight feet." Running at the somewhat slow speed of 30 knots, the Mark 18s would take about three minutes to reach the enemy ships. As the torpedoes were launched, the bridge lookout stated that he thought he had seen a pagoda mast on the target. Meanwhile, Reich had swung the *Sealion* around and fired three fish [torpedoes] at the second battleship in line. A minute later the crew saw the plumes and heard the explosions of the first salvo. Two of *Sealion*'s torpedoes had hit the battleship *Kongo*. Tully, who has extensively researched Japanese naval records, describes the damage inflicted by *Sealion*'s torpedoes:

> One in the port bow chain locker, the other aft of amidships, port side, under the No. 2 stack jarring the great battleship and causing spouts visible two ships astern on *Yamato*. The hits came with two loud booms followed by a "low grinding sound which vibrated the whole body of the ship." The bugle blared over the speakers, sounding the crew to action stations. Loudspeakers called for emergency teams to proceed to the inner anchor deck and effect shoring procedures. The torpedo hit there had torn a large gash in the bow. The second hit had flooded *Kongo*'s Nos. 6 and 8 boiler rooms, but the remaining boilers could provide adequate steam pressure, and despite the loss of fuel, enough remained to continue onward at fleet speed of 16 knots. . . . Though a considerable section of the port side and machinery spaces amidships aft were flooded, there was little initial concern among *Kongo*'s veteran crew.[5]

When the captain of *Nagato*, the second ship in line, saw the waterspouts from *Sealion*'s torpedoes exploding on the *Kongo*, he ordered the helm hard to port to comb any more torpedoes that might be coming from that direction. In doing so, *Nagato* deprived Reich of a hit on a second battleship. As the *Nagato* turned away, the *Sealion*'s second salvo went straight on to intersect the path of the first destroyer on the starboard side of the convoy, striking the destroyer *Urakaze*, which blew up a in brilliant circle of light, followed by a series of lesser explosions.[6]

After firing his second salvo from the stern tubes, Reich ordered the *Sealion* away from the convoy at flank speed until he thought it was safe enough to turn around and regain contact with the enemy. At 0310, *Sealion* had reloaded her torpedo tubes and was again tracking the convoy.

After receiving word that *Kongo*'s damage seemed manageable, Admiral Kurita, on board the flagship *Yamato*, decided to maintain speed in an attempt to escape pursuit by the American submarine. As the weather worsened and the ship continued to maintain 16 knots, the damage to *Kongo*'s hull got worse and brought on more flooding, forcing the battleship to suspend zigzagging and to slow down to 12 knots. Nevertheless, progressive flooding—spreading through leaks, fractured bulkheads, sprung seams, and pipes—continued throughout the ship. At this point, the convoy split into two groups with *Kongo* heading for Keelung, Formosa, for emergency repairs, escorted by the destroyers *Hamakaze* and *Isokaze*.[7]

Reich described what happened next: "We must have run about an hour [actually, it was two] in order to get back out and into a firing position and into what we thought was a good firing position and all at once the target slowed and stopped. We were out ahead about five or six thousand yards and if he'd have come by, we would have had him. But he sat there in the water—still. And then there was a tremendous explosion." It was so bright that Reich thought it looked "like a sunset at night." *Kongo* had succumbed to the flooding and her forward magazine exploded as the ship rolled over. Eli Reich had accomplished something no other submariner in the U.S. Navy had done before. He and his crew had sunk a battleship, an accomplishment for which he would receive his third Navy Cross—the second for the tonnage sunk on *Sealion*'s second patrol under his command.[8]

During the next few days. *Sealion* continued to patrol between China and Formosa, and then headed for Guam on October 28, 1944. When he arrived in Pearl Harbor, he was pleased to learn that signal intelligence had confirmed that the battleship that had gone down as a result of *Sealion*'s attack was the 37,000-ton *Kongo*. Reich was worn out. He had been continuously at sea for almost three years and had shouldered the responsibility of command for three war patrols. He was still expecting to go on a fourth patrol but was relieved by the powers that be in the Pacific submarine command and flown back to Pearl Harbor.[9]

After being detached from the *Sealion*, Reich was assigned duty as the assistant plans officer on the staff of the Commander, Submarine Force, Pacific Fleet. Part of his responsibility was to collect the intelligence obtained by the Intelligence Center at Pearl Harbor. Each day he would make his way up the hill to the Block House behind the Commander-in-Chief, Pacific Fleet's headquarters—where the Intelligence Center was housed—and he would collect the intelligence information relative to the submarine war in the Pacific. He would carry this information back down to the communications center in the submarine command's headquarters. In less than a half-hour from the time he had collected it, the information in the form of an ULTRA message would be sent by radio dispatch to the submarines in the Pacific. Reich also edited a monthly intelligence bulletin for the Pacific submarine force.[10]

In April 1944, Reich learned from his contemporaries that the destroyer force was looking for captains for their ships. By then Reich, having been a headquarters desk jockey for four months, was restless and wanted to get back to sea. He arranged a meeting with Cdr. Walker K. "Sol" Phillips, chief of staff, Destroyer Force, U.S. Pacific Fleet, to request a destroyer command. Reich explained that he had spent three years in four-stack destroyers and had submarine command experience. Although Commodore Phillips thought Reich would qualify for such an assignment, the submarine force, he was sure, would never release Reich. After "pulling a few strings" through his senior contacts in the submarine command, Reich received permission to be considered for a destroyer command and was subsequently ordered to the Prospective Commanding Officer's Training School in preparation for taking command of the destroyer *Compton* (DD 705). Reich entered the school, which was located not too far from Pearl Harbor, at the end of May. He spent ten days at the school, attending classes that covered the latest information and tactics involving radar, the combat information center, and gunnery.[11]

After completing the prospective commanding officer's training course, Reich was flown to Samar in the Philippines, where *Compton* was undergoing repairs. He arrived just as the destroyer was getting underway on July 4, 1945, to return to Okinawa for continued operations with the fleet. Enroute Reich relieved Cdr. Robert O. Strange and took over duty as the ship's captain. The

Compton, a 2,200-ton *Sumner*-class destroyer had twice the displacement of the four-stack *Clemson*-class destroyers that Reich had earlier served on, and it had considerably more armament. When the ship arrived in Buckner Bay, Okinawa, the kamikaze attacks were drawing to a close. Nevertheless, all ships maintained a gunnery watch, with gun directors and batteries manned. After escorting a convoy to Guam, *Compton* screened ships training in Leyte Gulf before returning to Buckner Bay. On August 12, 1944, Reich observed his first combat action as a surface warfare officer. At twilight, when the visibility was hazy and murky, a low-flying Japanese torpedo plane put a torpedo into the stern of the battleship *Pennsylvania* (BB 38) that was anchored less than 1,000 yards from the *Compton*.[12]

On August 25, *Compton* was designated as a courier ship and assigned to deliver operational orders, intelligence material, and mail to elements of the Fifth Fleet at sea. During this period *Compton*, with Reich conning the ship, made more than a dozen high-speed approaches on battleships and cruisers. As Reich later recalled, they "wouldn't slow down. They were operating at speeds in excess of 16 or 17 knots, and I'd come in about 20 knots on the stern, normally on the starboard quarter of a battleship and we'd pass a messenger line." They would send over a couple of mail bags and then be off to the next ship. On August 27, *Compton* collided with the battleship *Idaho* (BB 42) while attempting to transfer mail. Reich's ship was making about 22 knots when it got caught in the *Idaho's* propeller wash causing the portside forecastle to ride up on the battleship's armor belt. The impact put a twelve to fifteen-foot gash in the destroyer's thin plating and buckled a number of frames and plates on the port. The opening was six or eight feet above the waterline so there was not any immediate danger of flooding. *Compton* was immediately ordered to Sagami Bay forty miles southwest of Tokyo Bay to await the arrival of the tender *Piedmont* (AD 17) that would repair the damage. After the *Piedmont* arrived, but before repairs could begin, both ships were ordered into Tokyo Bay. On the morning of August 30, *Compton* proceeded into the harbor and dropped anchor, and in this "unofficial and impromptu fashion," she became the only Fifth Fleet ship to enter Tokyo Bay prior to the formal Japanese surrender.[13]

When the Japanese surrendered on the deck of the *Missouri* (BB 63) ending World War II, Reich was able to observe the proceedings from the *Compton* while alongside the *Piedmont*. Although the *Missouri* was across the bay, Reich could see it clearly. For the next three months, Reich commanded *Compton* as she served on patrol in the western Pacific.[14]

Toward the end of November 1945, Reich received orders to report to the Naval Academy. An assignment had come about as the result of a conversation that Reich had had in June with Rear Adm. Louis E. Denfeld, while both were in Pearl Harbor. Denfeld, who was commander of Battleship Division 9 at the time, was slated to become chief of the Bureau of Personnel on September 15, 1945, and he had already been thinking about the changes that would take place after the war ended. He told Reich that after the war ended, the Chief of Naval Operations (CNO) wanted the Naval Academy, whose wartime staff consisted mainly of reserve officers, to be staffed with commanding officers of submarines and destroyers, and the department heads of major ships. Denfeld asked Reich if he would be interested. "Well sure," answered Reich. Thus, on December 1, 1945, Reich boarded an airplane and made his way to Washington, D.C., where he was reunited with his wife who was living there with her parents.[15]

A day or two after he arrived in Washington, D.C., Reich received a call from the command duty officer. The officer explained that Reich was needed for the Chief of Naval Operations Tactical Publications Panel. The panel had been created by CNO Adm. Ernest J. King for the purpose of revising the Fleet Tactical Publications† based on the lessons learned during the war. According to Admiral King's dictate, only officers who had served on ships in the combat arenas could be assigned to the panel. Unfortunately for Reich, the submarine officer originally selected for this billet was no longer available. A replacement had to be found quickly, and Reich "fit the bill." Instead of reporting to the Naval Academy after his Christmas leave, Reich checked into the office of the CNO in the old Navy Building.[16]

---

† Fleet Tactical Publications were introduced in the 1920s to provide the foundation for training, to give the fleet standard operating procedures, and to serve as a point of departure for further development.

CHAPTER 6

FROM THE NAVAL ACADEMY TO TURKEY

IN AUGUST 1946, as the Chief of Naval Operations Tactical Publications Panel was wrapping up its work, Reich went to see the head of the panel, Capt. Alvin Duke Chandler. Reich wanted to get clearance and have the Bureau of Personnel issue orders for him to be sent to the Naval Academy. Although the panel's reports had been sent to the Government Printing Office, the galleys that were required for proofreading had not yet been completed. Reich told Chandler that he would finish the project while at the Naval Academy if Chandler would release him. Chandler agreed. Reich arrived at the Naval Academy in September and was assigned to the Department of Ordnance and Gunnery.[1]

The department was staffed similar to how it had been when Reich was a midshipman. It was all military except for a secretary or two. When he joined the department in September 1946, eighteen of the instructors were commanders, as opposed to the half-dozen lieutenant commanders, a flock of lieutenants, and a few JGs [junior grades] who had been there in Reich's day. This was the result of the policy established by CNO Admiral King to install instructors from the regular Navy with seagoing experience as fast as possible without regard for rank. Reich was not particularly happy with the quality of the students or with the way in which the curriculum was being taught. He did not think the attitude of many of the midshipmen was very good. In his opinion, "They weren't anxious to be in a combat organization," and he thought they probably wanted out of the Navy. He thought that they had been attracted to the Naval Academy because it was a nice place to be during the war since "they didn't have to join the Army or go out to sea." As for the curriculum, he thought it was very relaxed.[2]

In his day, the professors, all of whom were members of the Gun Club,* had been pretty smart. They expected the students to do all the work. "They might

* The Gun Club was an unofficial term used to describe the community of naval officers who specialized in gunnery, then considered the most important specialty in the Navy.

explain something to you, but more than anything else, they kept score on how you were doing." When Reich was a midshipman, his first-year ordnance course covered lots of ordnance equipment such as breech blocks, 5-inch gun recoil mechanisms, the makeup of the Ford Range Keeper, and so on. The textbook† they had used had illustrations and blueprint-like drawings to describe these materials, but the midshipmen had to figure out for themselves how they worked. And they would be quizzed on the material every day. "It wasn't an easy course," according to Reich. Now, in addition to a highly illustrated textbook, the instructors would bring in a full-scale mockup of the 5-inch gun mechanism, which they would then take apart and put back together. Reich felt that the Naval Academy was flooded with all these fancy devices that he considered to be "quick ways to learn." The instructors were supposed to grade their students every day. This created a problem for Reich because the teaching periods were only 55 minutes long. He would be explaining something and maybe get carried away, and suddenly 45 minutes had gone by, leaving little time for a quiz. Reich believed that "the profs were doing all of the work in terms of explaining this and trying to explain that, but . . . weren't touching the really hard stuff." The course that Reich took when he had been a midshipman "had lots of math in it, lots of sketching and describing, and lots of pressure on the individual midshipman to understand the mysteries of ordnance." In contrast to his experience, much of the difficult mathematics had now been eliminated, and the class was "a snap course" (which some claim is the same as it is today).[3]

Part of Reich's duties as an instructor was to mark his student's exams. Early on, he discovered that his students' punctuation, spelling, and penmanship were atrocious. He would be marking an exam in the committee room and start cursing. A fellow instructor sitting alongside would ask what he was hollering about? "Well," he would respond, "I don't know how much this kid knows about ordnance, but he doesn't know beans about punctuation. Not only that—he doesn't know how to spell, and one-third of this I can't even read. It's completely illegible and this guy's penmanship stinks."[4]

As we have seen before, Reich was never shy when it came to calling out failures in one of the Navy's programs or projects. All the instructors in the Department of Ordnance and Gunnery agreed that something was radically

† See, e.g., U.S. Naval Institute, *Naval Ordnance: A Textbook Prepared for Use of the Midshipmen of the United States Naval Academy* (Annapolis, MD: Naval Institute Press, 1933).

wrong. So, they got together and wrote a strong memorandum on the subject to the head of the Department of English and History. Not surprisingly, Reich helped put it together. The memo, in Reich's words, said in effect, "What the hell are you guys doing—here we're teaching the first class and second class, and these guys are about to go out and they don't know the rules of punctuation; they don't know this; and they don't know that, etc., etc. What in the hell is going on?"[5]

In early December 1947, Reich received a telephone call from Capt. Elton H. "Joe" Grenfell, the senior submarine officer in the Office of the Chief of Naval Operations. He was working to implement plans for a submarine exchange program with Turkey under the authority of the Truman Doctrine. His job was to see about the planning for the formation of the U.S. training group that was going to take the boats from the United States to Turkey and then conduct the training program. He had been checking the submarine files to see who was available and wanted to know if Reich would be interested. Reich, who was "getting a little itchy about being just a prof," said he was very interested and would be glad to be considered.[6]

While Reich was teaching ordnance and gunnery to the Naval Academy's midshipmen, the United States and the United Kingdom had been engaged in a bitter struggle to prevent the Soviet Union from dominating the European continent. In 1947, the Greek government was facing a serious communist insurgency that was supported by the Soviet Union and its Eastern European satellites. Turkey, while not directly threatened at the time, had been under varying degrees of Soviet pressure since 1945 and was therefore compelled to keep its entire army mobilized at great expense. President Harry S. Truman now faced a fateful decision. As he and Secretary of State George C. Marshall explained to congressional leaders on February 27, 1947, Greece was in grave danger of falling under communist control, thus isolating Turkey, which might then suffer a similar fate. Two weeks later, on March 12, 1947, President Truman addressed a joint session of Congress and asked for $400 million ($5.4 billion today) in military and economic assistance for Greece and Turkey—thus establishing a foreign policy aptly characterized as the Truman Doctrine. In addition to the funds, President Truman asked Congress to authorize a detail of American civilian and military personnel to Greece and Turkey, at the request of those countries, to assist in the tasks of reconstruction and to supervise the use of such financial and material assistance as would be furnished.

President Truman also recommended that authority also be provided for the instruction and training of selected Greek and Turkish personnel. The bill for aid to Greece and Turkey passed both houses of Congress by wide margins and was signed into law by Truman on May 22, 1947. The sanctioning of aid to Greece and Turkey by a Republican Congress marked the beginning of a long and enduring bipartisan Cold War foreign policy based on the Truman Doctrine, which declared, "It must be the policy of the United States to support free peoples who are resisting attempted subjugation by armed minorities or by outside pressures."[7]

The first step in this program was to dispatch a survey team of Army, Navy, Air Force, and State Department personnel to advise the U.S. ambassador to Turkey on the specific matériel requirements and organization of the Turkish armed forces, the training they should receive from the United States and the United Kingdom, and the nature and composition of any U.S. mission required in Turkey. The group, headed by Maj. Gen. Lunsford E. Oliver and Rear Adm. Ernest E. Herrmann, arrived in Ankara at the end of May 1947 and spent more than a month surveying the Turkish military situation. The survey group's recommendation, endorsed by the U.S. ambassador to Turkey, recommended separate Army, Navy, Air Force, and civilian missions to train the Turks in using and maintaining the U.S. equipment.[8]

A year earlier, the Turks had requested that four U.S. submarines be turned over to them. Although then CNO Adm. Chester W. Nimitz was sympathetic to the request, he believed that the Turkish navy was not up to handling the latest U.S. fleet submarines without additional training and logistic support, and he had opposed their request. Passage of the act authorizing U.S. assistance to Greece and Turkey enabled the Navy to provide the support deemed necessary by Nimitz, and plans were formulated to transfer four fleet boats to the Turkish navy.[9]

Just before Christmas 1946, Reich had received another call from Captain Grenfell to say that if Reich was still agreeable, he would recommend him for the mission as an ordnance and engineer officer. He said that things were moving fast, and that although all the policy decisions had been made, the actual detail planning had yet to be done, and he was forming a small group in his office to work on it and then proceed with the mission. Reich said yes, and he received orders after the first of the year to report immediately to Captain Grenfell's office in the old Navy Building.

The general plan, which was already underway when Reich reported to Captain Grenfell, was to train the personnel who were going to take over the submarines at the Submarine School in New London, Connecticut. Twelve officers and forty or fifty petty officers were already undergoing training when Reich arrived in New London, and the four submarines had already been selected for transfer from the Pacific Fleet. They sailed to New London by way of the Panama Canal and were not expected to arrive until the end of May 1947. The plan was to place four Turkish officers and ten petty officers and a reduced U.S. Navy crew of three or four officers and twenty men onboard each submarine, which would then be sailed from Connecticut to Turkey. Additional training of the Turks would be conducted while underway. Once in Turkey, there would be a turn-over ceremony and a change of flags, after which there would be a twenty-week training program supervised by the U.S. training group of sixteen to twenty officers and about one hundred enlisted men selected from the crews that had delivered the submarines.[10]

Reich's job was to make sure that the program received the logistic support necessary to sustain the training group that would remain after the boats were transferred to the Turkish navy. He worked on it from February through April with his supply officer, Lt. Cdr. Lenard Shea, and a physician in the Medical Corps, Lt. (jg) Halin G. Habib. Reich's first priority was to establish a facility to house the training complement. Fortuitously, the Navy had already worked out a guide that Reich could use in the form of a 150-page catalog prepared by the Seabees during World War II that spelled out all the equipment and components needed to create virtually any type of advanced naval base. Using the catalog, Reich selected the listing for a camp that was designed to accommodate up to 250 men. To Reich, it was "like an Erector Set,"‡ with a certain number of Quonset huts, the right quantity of plumbing supplies, all the electrical equipment needed, and so on. It showed what kind of water purification equipment was needed, what kind of equipment was necessary for a galley, what kinds of tools were needed, and so on, and it also identified the personnel needed to operate them. As Reich pointed out, it was one thing to pick an item out of a catalog and another thing to find out if it was available. To locate all the

‡ An Erector Set was a highly popular brand of metal toy construction sets originally patented by Alfred Carlton Gilbert and first sold by his company, the Mysto Manufacturing Company of New Haven, Connecticut, in 1913. It has many small parts that could be combined in numerous ways to create an infinite number of models.

TABLE 6.1

ADVANCED BASE COMPONENTS

N5C* camp buildings (250 men)—Northern

Designed to accompany an independent movement in a cold climate. It provides complete living facilities in huts for 25 officers, 200 men, and the 25-man complement below. Area Commanders should specify the number of officers and men each camp is to support if varying from the above numbers.

Personnel (approx.):

- No officers and 25 enlisted—Total 25

Material (major items only):

- Northern huts for housing, messing, recreation, and storage
- Northern hut for galley (including interior fixtures), dry and refrigerated storage
- Northern huts for laundry (with washing machine)
- Northern huts for shower—latrines (with chemical toilets)
- Water systems—Purification and distribution and distillation
- Power and lighting systems
- Transportation equipment—Trucks (for camp use only)
- Cobbler, barber, and tailor kits
- Cots, mattresses, bedding, and linens
- Galley gear and mess gear
- Special and protective clothing
- Consumables—30-day supply of dry provisions, 90-day supply of clothing and small stores, ships store stock, and standard stock items for approximately 250 men

Weight: Approx. 349 long tons

Cube: Approx. 642 measurement tons

* The N2C camp mentioned by Reich was the same as the NIC in reduced quantity for 100 men.

Source: U.S. Navy, Office of the Chief of Naval Operations, *Catalogue of Advanced Base Functional Components*, 3rd ed. (Washington, DC: Office of the Chief Naval Officer, 1945), 128.

material he needed, Reich journeyed to the big Seabee storage base in Quonset, Road Island, with Lieutenant Commander Shea. The Seabees on duty were more than happy to show the two officers what was available.[11]

Shea also knew that all kinds of surplus equipment that could be of use to them was coming on the market. He was also aware that a big ammunition depot just outside Boston was being dismantled and would likely be another source of the needed equipment. With Reich's permission, Shea made arrangements to visit the Boston site. He called Reich a couple of days later. "Gee whiz," he said, "there's all kinds of lumber, all kinds of plumbing equipment [wash bowls, urinals, etc.] and plumbing fittings—there's no end to it. And furthermore, the people here would be most happy if we could take it off their hands." The problem was how to get it to Turkey. Shea solved this problem by traveling to the Bayonne Naval Supply Depot in Bayonne, New Jersey, which was a major shipping and logistics center. Using his contacts, his understanding of the Bureau of Supplies procedures, and the authority of the CNO's exchange program, Shea was able to obtain bills of lading for trainloads of material that would be shipped by U.S. Army transports to Turkey. In addition to the material needed to build the training camp that Reich intended to set up in Turkey, he was also able—with Shea's help—to procure all the components needed to outfit a complete field torpedo workshop and an overhaul facility with a warhead magazine as well as a similar facility for mines, both of which were recommended by the Bureau of Ordnance based on the material described in the catalog of equipment for advanced bases.[12]

Reich realized that he would need an engineer/ordnance technical staff to support his duties as the project's ordnance and gunnery officer. To help recruit the men he needed, he traveled to the submarine base in New London in March 1947 to discuss his personnel needs with Commander "Sec" Johnson, the base's repair officer, whom he knew well. With Johnson's help, Reich put together an Engineer/Ordnance Group that consisted of four warrant officers—a chief machinist, a chief carpenter, a chief torpedoman, and a chief electrician—as well as a chief optical repairman, a welder, and a chief torpedo man, and so on. All the men had submarine backgrounds and had served either on submarines or on submarine tenders.[13]

In April 1947, Reich moved to the submarine base in New London and formed his division. The division left New London in early June and arrived at Izmir, Turkey, about ten days later. Izmir, which is on the Aegean Sea, was their first stop on the way to Golcuk, where the four boats were to be officially transferred to the Turks. Golcuk, located on the Gulf of Izmir, on the easternmost part of the Sea of Marmara, is about sixty or seventy miles from Istanbul. The plan as it had been originally formulated was to sail the boats through the Dardanelles while still under U.S. Navy command, flying the U.S. flag. This created a minor diplomatic problem: the Turks did not want them sailing into Turkish waters flying U.S. colors. The solution, which took a week or two to be work out, was to shift the flags in Izmir without a ceremony. The potential

Turkish skippers symbolically assumed command, but there was always an American officer on the bridge so that the U.S. Navy, in effect, would continue operating the boats until they arrived in Golcuk. Because of the delay, Reich and some of his supply and logistics people flew to Istanbul and from there went by car to Golcuk.[14]

When Reich arrived in Golcuk, he was received by the head of the U.S. Navy's engineering support activity, Capt. William M. Godson, CEC (Civil Engineering Corps). During their initial meeting, Reich learned that while Captain Godson had only a handful of men in his command, mostly Seabees, he had all types of heavy equipment from caterpillar tractors to diesel generators. Reich told Godson that his number one priority was to get his men ashore and get established. Godson agreed and said that he thought he could be helpful. Soon thereafter, Godson took Reich in a jeep tour around the area to determine the best place to establish the new camp. They found a perfect place behind the Turkish officer's club, on the beach on the Gulf of Izmir, three miles west of Godson's equipment yard.

"You know," Godson said, "we could put your camp right here."

Reich was amazed. To him it was a wilderness, but Godson assured him, "You've got these fellows who can drive these things, and I've got the equipment and we can do it." And that is what they did. From then on, Reich and Godson met every morning at the Turkish officer's club around breakfast time to discuss what had to be done that day. Reich would say to him, "I don't see how we're going to do all this." And Godson would reply, "I've got the brain and you've got the brawn with this crowd—we'll get the job done."[15]

Work on the camp, which was to be named Camp Lockwood after the renowned World War II commander of the U.S. Pacific Submarine Force, started in July, when they began to clear the land. With the land cleared, they put in sewer lines, drilled artesian wells, put in a water supply, and constructed a power plant. The Turks provided some assistance via the Turkish army, which contracted with the local civilians to provide bricklayers, laborers, carpenters, and so on.[16]

Once construction of Camp Lockwood was well underway, Reich turned his attention to the torpedo workshop and magazine. The Turks had elected to locate the facility eight or nine miles east of Golcuk in a place called Basiskele on the Gulf of Izmir. There was a Navy facility there with a dock, which was an important consideration because it was easier to move the equipment by water than by land. Reich's primary concern was making sure that all the components making up the workshop remained together and were delivered to where they were supposed to be installed. He sent his chief torpedoman to Basiskele to supervise the installation and would visit the site once a week to see how things were going. The mine overhaul and storage area was established and installed in a place called Diemendera, twelve miles or so west of Golcuk.[17]

Camp Lockwood opened sometime at the end of September or the beginning of October 1947 so that the American training crews, which had been billeted in the ex-German passenger ship SS *Tierhan*, could be relocated to better accommodations. Although Reich considered the *Tierhan* "a roach-infested ship," it was considered "fairly good accommodations" by Turkish standards. Needless to say, the American crews were very anxious to get off the ship, which put a great deal of pressure on Reich to complete Camp Lockwood as quickly as possible.[18]

In the meantime, the twenty-week training program was well underway. As the project's ordnance and gunnery officer, Reich was keenly interested in torpedo firing and in how well the Turks were handling the torpedoes. Reich, and the other officers overseeing the instruction of the Turkish crews, quickly found out that the Turks had different ideas about the work week than the Americans. According to the normal training schedule used by the Americans, the boats would be taken to sea first thing Monday morning and would come back at 1400 or 1500 on Friday. This was the training week that Reich recalled "was a great shock to the Turks," whose liberty began around noon on Thursday and did not end until Monday or even Tuesday morning. At first the Turks would not think of returning before then, but the U.S. commander of the expedition, Capt. Stanley P. Moseley, stood his ground and the Turks begrudgingly carried it out.[19]

Reich described the twenty-week program that ran from July 1948 to January 1949 in his oral history:

> To many of the Turks it was traumatic—because of this requirement to hit the [road] jack and go. They had these regular series of exercises and the whole idea of the thing was, initially, to wean the American officers off of the operations of the boats and over to the Turks. Now, the first part of the program was pretty much oriented towards the safety of the boat and making sure that the people knew the right things to do to

> maintain safety and then to get into the advanced phases of training, i.e., torpedo shooting, and all the other things that one does in submarine warfare.[20]

The Americans also had to bring the Turkish crews up to the required level of maintenance. By the time the training had ended in January, everybody was happy. The Turks felt that they had been put upon, that they had had to work their fannies off, and that the Americans were saying nice things about them.[21]

After the official Turkish training period ended, the group of American instructors was reduced from twenty-five officers to six or seven men whose mission was to act as advisers. They stayed on for another six months while the rest of the personnel returned to the United States. Things seemed to be going well until April 1949, when the Turkish General Staff began complaining about the performance of the American torpedoes, which they claimed were not running properly. The Turks believed that they had been given shoddy goods and set up a Board of Inquiry at Golcuk to investigate the situation. Reich had not been involved in this brouhaha until Captain Moseley ordered him to investigate what was going on behalf of the U.S. Navy. By then Reich's Engineering/Ordnance Group had constructed and outfitted the torpedo workshop and the warhead magazine. They had also established a torpedo school, which was supervised by Reich's chief torpedoman. The school provided hands-on training for senior petty officers and torpedo officers in the Turkish Navy.[22]

Reich and his torpedo people had not paid much attention to the operations conducted by the Turks during the previous three months; now they had to find out what had been going on. It turned out that while the Turks had been on their own, more than half of the Mark 14 torpedoes had run erratically, did not run at all, or had missed their targets. With the cooperation of the Turkish Navy, Reich was able to place his people on the transferred submarines to observe what was going on. It took only a few days for him to discover that as soon as the Navy's trainers left, the people handling the torpedoes were no longer the highly skilled senior petty officers but conscripted sailors known as "askri." Many of them could not even read, and these were the people who were checking the torpedoes and making adjustments.[23]

With the help of Rear Admiral Gemel Bozark, the Turkish admiral in charge of the Golcuk Navy Yard, Reich requested and received permission to conduct a series of controlled firings of torpedoes that had been prepared under U.S. supervision. Reich and another U.S. Navy officer were allowed to observe the torpedoes that were fired into nets to capture their points of entry and the torpedoes' depths. A crosshead had been painted right in the middle of the barge-mounted nets so that there would be no question of the aiming point. After Reich's people made sure that all the torpedoes' preliminary and final preparations had been properly conducted, the torpedoes were fired from 800 to 1,000 yards from the target net. The firings took place with the submarine on the surface, but trimmed down. This provided a stable platform in a calm sea. The idea was to hold constant as many variables as possible in order to obtain a reliable measure of the performance of the torpedo's gyro and depth setting mechanism during a normal torpedo run.[24]

The Turkish captains were a proud lot who clearly resented the intrusion of the U.S. observers, and they were not easily persuaded to take instruction from these foreigners. So Reich let them fire the first torpedo without saying a word, and he watched the wake of the torpedo as it missed the aiming point by 150 yards. Reich insisted upon watching through the aft search periscope as the next shot was being taken. As Reich looked on, the Turkish captain, using the forward attack periscope, attempted to line up crosshairs on the target. Just as he gave the order to fire, Reich screamed, "No!" But it was too late; the captain had not taken into account the submarine's movement and missed again. According to Reich, it took "a little bit of saavy [*sic*]" to get it just right. "Let's be on the target when we fire," explained Reich. With Reich looking over the captain's shoulder, the next twenty-three torpedoes hit the target net within three to eight feet of their depth setting. The firing tests proved there was nothing wrong with the torpedoes, and it reinforced Reich's concept of handling complex ordnance. He later said, "There is always something wrong with complex ordnance if it is not properly maintained, not properly adjusted and not properly used." The ordnance would not do it by itself. It needed a lot of attention and skill to keep it working all the time.[25]

Reich left Turkey in July 1949 with orders to attend the fall class at the Armed Forces Staff College in Norfolk, Virginia, with an interim assignment to the Royal Navy's Anti-Submarine Warfare (ASW) school in Londonderry, Northern Ireland. The former was due to the interest of Captain Grenfell, who had nominated him for the course.[26]

CHAPTER 7

FROM STAFF OFFICER TO DESTROYER COMMAND

THE ARMED FORCES STAFF COLLEGE that Eli Reich entered in September 1949 had been established at Norfolk, Virginia, in the fall of 1947 to train selected officers of the armed forces in joint operations. Like Reich, 80 percent of the class was made up of commanders such as himself with an average of fourteen years of service. Because of World War II, the military backgrounds of the first few classes were based more on combat experience than on the completion of formal instruction in the lower military schools. Entrance requirements were frequently waved for officers like Reich who had exceptional war records.[1]

The students in Reich's class, the sixth held at the school, came from all branches of the services and included a group of officers from the Canadian and Royal navies. Reich got to know many of the men from the other services, in part due to the mixed housing arrangements that placed an army, navy, and air force (or foreigner) in the same room. During the five-month course, the students were exposed to the following scope of instruction:

1. Orientation in organization, characteristics, and employment of the army, naval, and air forces, and the relation of those forces to each other.
2. Study of joint staff techniques and procedures.
3. Preparation of plans for amphibious and airborne operations involving the employment of joint forces.
4. Study of the organization, composition, and functions of theaters of operation and major task forces, and the responsibilities (strategical, tactical, and logistic) of the commanders thereof.
5. Study of the trend of new weapons and scientific developments, and their effect upon joint operations.

Reich considered it to be a "very fine school," and he was particularly impressed with the opportunity it offered to associate with officers from the other services.

Based on his experience at the school, Reich came away with the "feeling that military life was the same the whole world over, it didn't make much difference what service you happened to be in."[2]

While still at the school, Reich received orders to the staff of the Commander, Submarines, Pacific Fleet (ComSubPac), as training officer. When he arrived at the command's headquarters in Honolulu, Hawaii, Reich found that as training officer he also had duty as the assistant to the readiness officer, Capt. William D. Irvin. After he had been on this assignment for a few weeks he was directed to serve as understudy to Cdr. Antone R. "Tony" Gallaher, whose submarine division was carrying out the KAYO project. The project was initiated by Chief of Naval Operations Adm. Forrest P. Sherman in January 1949, when he directed the Atlantic and Pacific Fleets to create a submarine division to develop techniques for submarines to detect and destroy enemy submarines. The centerpiece of the project soon became a new type of submarine, designated the SSK. It would lie in wait to ambush enemy submarines off Soviet ports and in channels and straits where Soviet submarines would transit—on the surface or snorkeling—enroute to and from the Atlantic shipping lanes. A key component of the SSK design was the BQR-4 passive sonar, the first sonar array developed by the U.S. Navy consisting of fifty-eight hydrophones, each ten feet high, mounted in a circular arrangement on the bridge fairwater. When Reich was assigned to the project in the spring of 1950, the first of these specialized hunter-killer submarines, the *Barracuda* (SSK 1), was still under construction at Electric Boat division of General Dynamics Corporation in Groton, Connecticut.[3]

After Reich had been with the project for about a month, the powers that be within the hierarchy of the Navy decided to phase out the project in the Pacific and to concentrate the antisubmarine warfare (ASW) effort in the Atlantic instead. Reich believed that this decision was made because "the state-of-the-art in ASW was greater in the Atlantic, primarily because they were closer to hardware developments and the interest of the East Coast laboratories." As the acting division commander, Reich had the task of closing out the division's role in the KAYO project, which involved completing certain studies and providing final reports on the work that had been done in the Pacific.[4]

Reich's disengagement from the KAYO project may not have been a bad thing, given the fact that the three K-boats [SSK submarines] completed between 1951 and 1952 were cramped and uncomfortable, and their slow transit speed limited their ability to be moved into forward areas during a crisis. They also

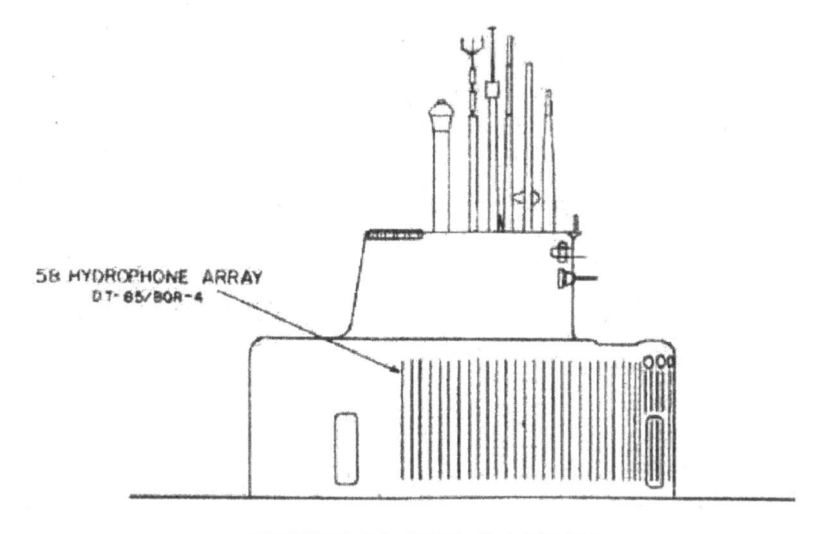

FIGURE 7.1. BQR-4 ARRAY

Source: "U.S. Navy Hunter-Killer Submarines," *Weapons and Warfare*, accessed August 30, 2022, https://weaponsandwarfare.com/2017/01/25/us-navy-hunter-killer-submarines/

"suffered from having diesel engines that were difficult to maintain, an unreliable and insufficient freshwater plant, undependable electrical generators, and slow speed." As Gary Weir has noted, regarding the naval-industrial complex and the American submarine, "Rather than continuing with the smaller K series, the Navy spent the balance of the decade fitting seven old *Gato*-class fleet submarines with the more advanced BQR-4 sonar to take over the hunter-killer mission."[5]

After completing the final KAYO reports around the end of May or the beginning of July 1950, Reich returned to duty as ComSubPac's training officer, relieving Cdr. Robert A. Keating Jr. Keating informed Reich of the big project underway to revise all the training exercises that had begun under Captain Irvin's predecessor. The project was half done, and Keating was turning it over to Reich to complete. Reich looked over the drafts of the exercises that were still in progress and had several discussions with Captain Irvin, his immediate superior. He spent a lot of time on the waterfront and went out as an observer on various boats in order to obtain first-hand information on the status of ComSubPac's submarine training program. After a month or two, Reich concluded that the duty on the submarines based at Pearl Harbor "had tuned into a country club type of living. . . . All the guts had been taken [out] of the rough training exercises" that had been the norm during World War II.[6]

Instead of an intense five- or six-day training exercise involving numerous torpedo attacks and exercises conducted by ASW defending forces, the boats would spend just one day at sea. Reich found

> that some of the exercise war patrols being run in Pearl [Harbor] (that were supposed to get underway at daylight) might start . . . officially at 6:30 in the morning, but the ship [submarine] never left the dock. They buttoned the boat up and they would simulate submergence alongside the dock. They would get through certain exercises right there at the harbor and then, instead of it being a grueling five- or ten-day affair, it would turn out to be one day at sea—out in the morning and then come back late in the evening and that was it. Well, to a couple of crusty guys like Bill Irvin and myself, we thought this was 'going to hell in a hand basket.'

After discussing the issue with Captain Irvin, Reich was instructed to rewrite all the training instructions to bring them closer to what had been in the syllabus for training during the latter part of World War II. Although this was a strenuous undertaking, Reich approached the task with zest. When he had completed all the revisions, Reich brought all the drafts that were ready to Captain Irvin. After discussing some of the finer points, Irvin told Reich, "That's great," indicating that he was very pleased with what Reich had prepared.[7]

Then all hell broke loose. It seems that Captain Grenfell, Rear Adm. John H. "Babe" Brown's chief of staff, was not aware of what was going on. He knew that revisions were in the works, but he did not know the details. Reich was working at his desk finalizing the revisions one Saturday morning in October 1950, when he answered a call from Captain Grenfell. The first thing the chief of staff wanted to know was where Reich's boss was. Reich told him that Captain Irvin had taken the day off and was on the north side of the island.

"Well, you get down here," he declared.

So, Reich went to Grenfell's office and was inundated with a series of questions:

"What are you and Bill Irvin up to?'

"What are you guys trying to do?"

"What's this that I hear from the waterfront?"

"What's this I hear from the squadron commanders?"

Reich was taken aback but quickly began to understand what was going on. Before he could say anything, Grenfell said to him, "Listen, I want you to get word to your boss to see me as soon as you can locate him." Later that evening, Reich stopped by Captain Irvin's house to tell him what had taken place. "OK," he said, "I'll take care of it."[8]

First thing on Monday morning, Reich got called into Captain Irvin's office.

"Did the chief of staff say, 'this and so?'" Irvin asked.

"Yes, Sir, that's exactly what he said," Reich replied.

Irvin responded with a litany of what was wrong with the new revisions for the training instructions "making damn clear" to Reich that they were on the wrong track. Then, he told Reich to find the originals of the previous instructions and to re-date them so that the new revision would essentially be the same as before.[9]

When the two men got together later that day Reich was shocked to hear what Captain Irvin had to say.

"Look," he said, "we're not good for this outfit and this outfit isn't good for us. It's obvious that we are out of step with how people want things done around here. We've got our instructions and what to do and that's what we're going to do. I don't agree in principle with it."

"Well, damned if I do," Reich answered.

"Well," stated Irvin, "I'm going to take steps to find other duty."

That was all that Reich needed to make up his own mind. The Korean War had started, and he was aware that the Navy was pulling a lot of ships out of reserve. Reich's last sea job had been the command of a destroyer, and he had had two years of duty in four-stackers. So he sat down and wrote a letter to the Bureau of Personnel requesting duty in command of a destroyer. He wanted to get out of the submarine community and thought that there was a good chance he would be acceptable as destroyer skipper.[10]

Reich's request needed to go up the chain of command, so he first submitted it to Captain Irvin, his immediate supervisor. Irvin favorably endorsed it and sent it to Chief of Staff Grenfell, who sent for Reich. Reich had known Grenfell for some time, and they got along fine; but Reich did not always agree with him. Grenfell wanted to know why Reich, who was now in line for command of a submarine division, wanted a transfer. Grenfell vehemently opposed Reich's decision but agreed to forward it to Admiral Brown.[11]

A few days later, Admiral Brown sent for Reich, and the two of them had a frank discussion over a cup of coffee. Look, Brown said, "I know what you have in mind to do, and I applaud it. I am going to endorse this favorably and you are going to leave Pearl with my blessings." Admiral Brown was as good as his word and put in a favorable endorsement; he even gave Reich an excellent fitness report.[12]

Just before Christmas 1950, Reich received orders to proceed to Charleston, South Carolina, and report to the Charleston Group, Atlantic Fleet Reserve, as the prospective commanding officer (PCO) of the *Stoddard* (DD 566), a World War II *Fletcher*-class destroyer moored in the Cooper River that was to be taken out of "mothballs" and placed back in commission. When he arrived in mid-February 1951, Reich found "the place was a beehive of activity," and he was given a brochure-like document detailing what the schedule was, how the crew would be assembled, and the procedures that were to be taken during the reactivation process. Reich was pleased to be in Charleston, knowing that he would soon be getting a ship. His immediate concern was rounding up the officers and men being ordered in and getting them organized. He was not concerned with the reactivation itself, because that would be taken care of by the reserve fleet. He had been told that they had a tried-and-true procedure for getting the mothballed ships into shape.[13]

Recommissioning a ship was a difficult and challenging job for a PCO. To begin with, the material situation on the ship was questionable. The PCO had to ensure it was safe to operate, especially the boiler and the main propulsion plant. And it had to be habitable. And if it were a fighting ship, as the *Stoddard* was, the PCO had to worry about fire control, the gun mounts, the directors, and so on. And then there was the problem of manning and organizing the crew. And the PCO had to be doing all of this at the same time. Reich considered it "a hell of a strain."[14]

Reich had not been in Charleston very long before he began hearing horror stories from his contemporaries who were either skippers or PCOs. Hypothetically, they were supposed to be steaming down the river heading to the open seas around thirty days after they arrived for commissioning. But according to the PCOs and skippers who had been there when Reich arrived, the "reserve fleet was just for show, but not so good when it came to putting up and putting out." Although the Bureau of Ships had laid out a very impressive scheme for mothballing these ships, the reality was quite different. It takes competent, well supervised people to properly lay up a ship. Reich knew this from his experience decommissioning the *Gilmer* in 1938. But the men who were supposed to have been decommissioning the *Stoddard* and the fifty other destroyers in the

Cooper River just wanted to go home—to leave the Navy for their long-awaited return to civilian life. And because the Navy was closing out and terminating contracts for the convenience of the government, the material situation was in chaos. All this led to considerable degradation of systems and equipment on the decommissioned ships that the reserve fleet had not adequately addressed.[15]

One of the commanding officers of the newly recommissioned ships to whom Reich talked, told of the horrendous problems he had faced in trying to deal with the people in charge of the reserve fleet. He did not agree with the procedures they were using to bring the powerplant up to snuff. He was shrewd enough, to document his concerns in writing, stating that he was unhappy with the work procedures that had been used to recondition his ship and was not sure what would occur once they got to sea. Sure enough, one of the boilers failed during the sea trials, and it took another month to get it repaired.[16]

Thus, Reich was well informed about what to expect from the work being carried out by the Reserve Fleet Command. From other officers he also learned a valuable piece of information. If possible, he was advised, he should try to locate the Bureau of Inspection and Survey's last report on the ship. Although Reich was not able to locate one for the *Stoddard*, he was able to get ahold of four or five Board of Inspection and Survey "INSUR" reports for other *Fletcher*-class destroyers that had been mothballed in the Cooper River. Reich found that the comments on the deficiencies listed in the reports for ships comparable to the *Stoddard* were pretty close to what he was observing on his own ship. By then he had obtained a commissioning crew of 18 officers and around 250 men. About 85 percent of the crew were reservists who had been recalled to duty as a result of the Korean War. They were not too happy about it, either. Using the comments from the INSUR reports as a guide, Reich divided the ship into a number of spaces, each of which the crew then carefully inspected. As the crew went through the various spaces assigned to them, they found similar examples of what had been in the reports and confirmed the deficiencies listed by the INSURs. In some cases, the problem was even worse than what had been described in the reports. When Reich tried to bring this information to the powers that be within the reserve fleet, they hurriedly brushed him off with the explanation that he "was standing in the way of them carrying out their tried and proven activation procedure, and they weren't about to look into this or that." Reich, mindful of what others had told him and realizing that a frontal approach was not going to pay off, began to document not only the results of his inspections but his efforts to bring them to the attention of the reserve fleet.[17]

As they were nearing the day for the commissioning ceremony, the commander of the Charleston Group came down to look over the *Stoddard*. After looking the ship over, he sent for Reich to publicly admonish him for not cooperating with his people or being mindful of the group's requirements. He was upset to no end (according to Reich) that the topsides had not been painted and the waterline had not been cut in. He was not going to stand for it, and in the morning the reserve fleet spray painters were coming down and they were going to paint the ship. As Reich had been forewarned, the service fleet emphasized the importance of paintwork and how good a ship looked at commissioning. The next day, as promised, the paint crew arrived and painted the ship. "If you stood a thousand yards away from that ship," Reich later declared, "it must have looked a beauty, but I was disgusted."[18]

On the *Stoddard*'s recommissioning day, March 9, 1951, various minor officials made little speeches before Reich read the orders making him commanding officer, but inside he was seething. He was so angry that he refused to sign the one-page pro forma document that the Charleston Group had prepared indicating that he had formally accepted the ship for the Navy and was now responsible for it. As Reich would do again in the future when he was unhappy with the material conditions within his command, he wrote a report. The four-page letter with its ten- or twelve-page appendixes fully documented the material deficiencies that had not been fixed and his reasons for not accepting the ship. When word that Reich was not going to sign the pro forma acceptance letter reached Capt. Robert K. "Blinn" VanMater, the representative of the commander of the Destroyer Force, Atlantic Fleet (DesLant), asked to see him. VanMater had been the intelligence officer at the American embassy in Turkey when Reich had been stationed there, and the two had a passing relationship. It was obvious to Reich that VanMater did not know what was going on and that he wanted to know why Reich was rocking the boat. Reich explained the situation to VanMater and asked if he (VanMater) would take the *Stoddard* under such circumstances. Reich made things even clearer by repeating a few things that were said in Navy regulations that pertained to his situation. VanMater was very understanding, stating that it was OK, and that maybe that was the way Reich had to do it. So, Reich accepted the ship with reservations. Instead of getting underway in thirty days, it was closer to sixty.[19]

Reich's approach to handling the problem posed by *Stoddard*'s material deficiencies, according to one veteran naval officer, is what made him stand out from among his peers. It involved a significant career risk, but this course of action would serve him well later on, when he was given command of the guided missile cruiser *Canberra* (CAG 2).[20]

The *Stoddard*, with Reich in command, departed Cooper River on April 1, 1951, stopped in Norfolk for a few days, and then steamed to its new home port of Newport, Road Island, which was home to DesLant's headquarters. After arriving in Newport, Reich spent ten days supervising training before heading for Guantanamo Bay and the training center there. Reich considered it a great training area: you could start shooting, conduct air exercises, or fire torpedoes within a mile of clearing the entrance to the bay. The weather was good, and there was excellent service in terms of target rafts and aiming points on the beach for shore bombardment. The training was intense for a crew that consisted of approximately 5 percent experienced regulars, 15 percent reservists, 60 percent Naval Academy and service school graduates, and 20 percent recruits fresh out of boot camp. "All of the mistakes that were anticipated were made," according to one member of the crew. "After two weeks of shakedown and training, a rhythm set in. When general quarters sounded, every crew member knew where to go, how to get there, and what to do upon arrival. 'Manned and Ready' was reported on time or earlier than expected. Guns were loaded, targets acquired, and firings occurred much earlier than before. Communications were faster with no voice hesitation, reaction to commands were automatic."[21]

After the training was completed, *Stoddard* operated along the Atlantic Seaboard until the fall, when she was sent to the Boston Navy Yard for overhaul and modernization. All her World War II vintage 40-mm anti-aircraft guns were removed and replaced with twin 3-inch/50-caliber rapid fire anti-aircraft guns. Although the 3-inch/50-caliber gun had a slower rate of fire than the 40-mm guns it replaced, the 3-inch projectile, packed with three-quarters of a pound of high explosives, was five times as powerful. Moreover, the 3-inch projectile was the smallest that could be fitted with the VT proximity fuze. Unlike the 40-mm, which relied on a human-operated director, the 3-inch/50-caliber gun was automatically radar controlled by a Mark 56 director installed with each mount. Together these improvements provided a system that was more likely to result in an aircraft kill in the days before guided missiles became the main anti-aircraft weapon on the Navy's ships. In addition to the replacement of her 40-mm guns, *Stoddard*'s forward torpedo tubes were removed, and most of the steel structure in the 01 level amidships and aft was replaced with aluminum to reduce weight.[22]

After the refit was complete, *Stoddard* steamed back to Guantanamo for shakedown and more training on the newly installed weapons. By the end of March 1952, the ship returned to her home port at Newport. In May, *Stoddard* was tied up alongside the destroyer tender *Yosemite* (AD 19) for a two-week in-port period, making ready to deploy to the Mediterranean with the Sixth Fleet. Reich was looking forward to deploying. He had finally gotten the ship into a reasonable material condition, and the crew was fairly well trained. Reich was confident and ready to go when he was invited to have lunch with DesLant's chief of staff, Capt. Alfred G. "Corky" Ward. During lunch, Captain Ward asked if Reich had seen his orders.

"No," replied Reich.

"It seems to me I read a dispatch this morning," said the captain.

"Well," Reich said, "maybe it's down in the ship."

After lunch Reich walked down to his ship, and as he boarded the watch officer told him that Cdr. John Baumeister Jr. was here as his relief. Reich went down to the wardroom and found Baumeister. "I'm going to relieve you," Baumeister said. "I was ordered here in a hurry." Within an hour or so Reich received an official dispatch ordering him to detach immediately and proceed directly to the Bureau of Ordnance in Washington, D.C., for duty as head of the Torpedo Research Branch.[23]

CHAPTER 8

FROM THE TORPEDO RESEARCH BRANCH TO THE INDUSTRIAL WAR COLLEGE

WHEN HE REPORTED TO THE BUREAU OF ORDNANCE in May 1952, Reich's boss, Capt. Maurice Kelly, the director of research, told Reich that there were things that needed to be done in the branch and that he was glad to see him. He gave Reich the impression that it was just as well he did not have a background in the Bureau of Ordnance's way of doing things. Kelly assured Reich that he had his full support and a free hand to do what was necessary. The message Reich got was that Kelly wanted him to clean house.[1]

Reich never revealed who had selected him for this assignment—assuming he ever discovered it himself. It is wishful thinking to believe that someone remembered his wartime experience with the development of the Mark 18 torpedo; or perhaps Reich was selected because of the work he did in Turkey. More likely it was simply that Reich was a well-respected accomplished submariner who happened to be available. Reich may not have been thrilled with the assignment, but the experience would be of immense value later in his career.

Within a day or two of reporting in, Reich decided to touch base with the submarine people he knew in the Office of the Chief of Naval Operations (OpNav) to see if he could find out why he had been selected to take over the Torpedo Research Branch. Reich soon discovered that the people in OpNav were very unhappy with Bureau of Ordnance's handling of torpedo research and that they were highly critical of Dr. Frederick Maxfield, a civilian employed by the branch as technical director. They told Reich that "there was a certain degree of backbiting going on due to the incompetence, or the independence, of certain civilians—both in the Bureau of Ordnance and its support laboratories—about the manner in which they wanted to go forward with torpedo research and development."[2]

Reich spent the first two weeks on the job sitting next to Cdr. Robert D. Risser, whom he was about to relieve, in order to observe what was going on in the branch. It quickly became clear to Reich that research and development in the branch was completely controlled, both financially and technically, by Dr. Maxfield. Reich liked Risser, but it was obvious that Risser was letting Maxfield run the show. There was tension among the other civilians in the branch, too, because "Maxfield ran the roost like a Prussian." After relieving Commander Risser, Reich had a long discussion with Cdr. Richard D. "Dick" Mugg, the number two military person in the branch. Mugg was very unhappy with the way things had been going for the past two years. He did not like the fact that everything was being determined by Dr. Maxwell. Next, Reich sat down and reviewed the job description he was given, which gave him the authority to run the branch. Right then and there he determined that he, and not Dr. Maxwell, was going to be in charge.

After a week or two had passed, Reich had a formal, one-on-one meeting with Dr. Maxfield. During the meeting, Reich, having reviewed his position and what the branch head was supposed to be doing, explained how he, Reich, intended to carry out his assignment as it was laid out. The first thing he was going to do, he told Maxfield, was to take over the actual operations budget. Maxfield was henceforth to act primarily as a technical adviser. Maxwell was very taken aback by Reich's pronouncement. He had made the technical and financial decisions for the previous four or five years and felt that while naval officers assigned to the Bureau came and went, he was the continuity that kept the show on the road. According to Reich, the meeting, which lasted an hour or two, "really shook him up." Maxfield resigned within six months and left the Bureau.[3]

With Maxfield gone, Reich was able to take over the chairmanship of the Torpedo Advisory Committee (TORPAC), which was made up of the technical directors of the principal laboratories that worked in the underwater and torpedo field. Maxfield had dominated this group, too, and the laboratory directors were unanimous in their general dislike of the man. Reich considered the committee to be very important to torpedo research, but in his words, it "needed to be shepherded, needed to be led, needed to be beat upon, and needed to be directed." As head of the Torpedo Research Branch, Reich was responsible for dishing out close to $25 million annually to these laboratories. This was a considerable amount of money, equivalent to over $1 billion today.

The branch was solely supporting some of these labs, and Reich believed that the branch should get something tangible out of it.[4]

Antisubmarine warfare was a major mission for the U.S. Navy when Reich was assigned to the Bureau of Ordnance. Two post–World War II developed torpedoes designed to be used against submarines were in production when Reich assumed the leadership of the Torpedo Research Branch. The first of these to enter service was the Mark 35, which the General Electric Company began producing in 1949. It was an outgrowth of the Mark 24 (FIDO) homing torpedo developed in 1944. The electrically powered, 1,690-pound Mark 35 was a deep-diving, long-range homing torpedo intended as a "universal" torpedo that could be launched from any platform (although the air-drop feature had been eliminated in 1947). The other major torpedo program centered around the Mark 43 Mod 1 torpedo that could be deployed from fixed wing or rotary aircraft. The Mark 43 Mod 1, like the Mark 35 that preceded it, had an active sonar system. When it was fired, the Mark 45 ran straight for an adjustable distance that could be preset for 550, 1,000, or 2,000 yards). It then began a search pattern using its sonar, which could detect a target (an enemy submarine), up to 500 yards distant, and it would then home in it. Reich considered it a pretty good torpedo, despite questions of reliability and repeatability that were characteristic of the technology inherent in these complex designs.[5]

When Reich took over the branch, he arranged to visit the GE plant in Pittsfield, Massachusetts, that was producing the Mark 35 for the U.S. Navy. It was an expensive torpedo that the Navy had invested a lot of money into. During his tour of the plant, Reich was amazed to find close to three hundred torpedoes in various stages of production. He was surprised to learn that although GE had produced all the parts for the torpedo, the company's engineers were having great difficulty in getting the highly complex homing system in the nose to meet the required production specifications. No two of the homing devices in the nose cones looked the same to Reich, who got the impression that they were using "laboratory methods" (each homing device made individually, one at a time) for production.[6]

The nose cone contained the highly sensitive acoustic sensors that were used to detect the target, the electronics needed for homing, and the sonar transmitter. In all likelihood, it was similar to the system that GE had used on the Mark 32, the first active acoustic torpedo introduced in the U.S. Navy. The Mark

32's detection system used a magnetostrive transducer* that was four elements wide and eight elements high, split into an upper and a lower half. Homing signals in the vertical plane, derived from the phase differences of the acoustic signals returned from the target submarine's hull (which was only 25 feet high), were compared by two halves of the transducer to provide proportional control in the vertical plane, where the target was smaller. A simpler on-off control mechanism was used in the horizontal plane, where the target was larger. The control system was a very complex mechanism that had many adjustments that could be made for various types of tactical uses. The nose cone also contained the sonar transmitter that relied on two vacuum tubes to power the transmitter's transducer. The combination of vacuum tube technology using individual electronic components assembled on a breadboard created hordes of reliability problems.[7]

After reviewing the program for close to a year, Reich concluded that the torpedo was a "monster" that had been prematurely committed to production and was not fit for fleet use. After a very acrimonious review meeting with some of GE's engineers and with certain people in the Bureau of Ordnance and OpNav, Reich spearheaded cancellation of the program. By then four hundred Mark 35 torpedoes had been produced. A lesson was learned: technology readiness must be confirmed before production can begin. The Mark 35 saw limited use and was withdrawn from service in 1960.[8]

While Reich was in charge of the Torpedo Research Branch, Westinghouse and the Ordnance Research Laboratory at Pennsylvania State College, which functioned as Bureau of Ordnance's technical representative, teamed up to produce the radically new Mark 37 submarine-launched torpedo. The Mark 37 was a wire-guided torpedo that the engineers at the Naval Underwater Ordnance Station† at Newport, Rhode Island, had come up with to overcome the tactical limitations of the Mark 35. Wire guidance enabled the launching submarine to control the torpedo's initial maneuvers through a wire unreeling from both the submarine and the torpedo. This feature permitted initial guidance by the more-capable submarine sonar rather than by that of the torpedo.[9]

---

* A type of transducer that relies on magnetic material to convert mechanical energy into electrical energy.

† The Underwater Ordnance Station replaced the Naval Torpedo Station in 1951, when the Torpedo Station was permanently disestablished. Thereafter, the manufacture of torpedoes was awarded to private industry.

After reading through a number of technical and management reviews of the Mark 37 program, Reich found that it had the same problems as the Mark 35. Fortunately, the Mark 37 program was still in the early stages of development and no commitment to production had been made. Westinghouse and certain members of the Bureau of Ordnance's staff were anxious to begin production, but Reich was determined that he "was not going to be party to the kind of mistakes that had been made on the 35 and to push the damn thing into production before we knew what the state of the design really was." Reich, against the advice of both Westinghouse and the Ordnance Research Laboratory, decided to conduct an engineering and technical audit of the design. To accomplish this task, he had the bureau issue a contract to the C. W. Smith firm, a consulting engineering company based in Chicago, and set them up in the Naval Ordnance Plant in Forest Park, Illinois, that had produced torpedoes during World War II.

Once the staff supplied by C. W. Smith were established in the Naval Ordnance Plant, Reich directed Westinghouse to send all the blueprints, drawings, test specifications, and whatever technical data was used to support the design of the Mark 37 to the plant in Forest Park. As soon as the engineering review was finished, Reich intended to establish a pilot production line at the Forest Park plant. Neither Westinghouse nor the Ordnance Research Laboratory was very happy about this. The latter thought their stewardship was being questioned, which it was. Reich knew the folks in the Ordnance Research Laboratory were not production engineers: they were people in an academic laboratory. They were very good at laboratory research, but they were not knowledgeable in what Reich called the "nitty-gritty" of the production floor. As for Westinghouse, they considered the Mark 37 design complete and had expected to begin production.[10]

Within a month of starting their review of the Mark 37, the independent engineers from C. W. Smith were uncovering glaring problems with Westinghouse's design. The engineers hired by Reich were manufacturing specialists who were very knowledgeable in the production processes and the machinery necessary to fabricate the highly complex parts of the Mark 37 torpedo. With the help of C. W. Smith, together with contracts issued to Westinghouse and the Naval Ordnance Laboratory at Pennsylvania State College for technical support, the pilot production run of the Mark 37 was successfully completed at the at Naval Ordnance Plant. The Mark 37 went on to full production and became one of the most prolific torpedoes fielded by the U.S. Navy.[11]

On April 1, 1954, Eli Reich was promoted to captain and moved up to relieve the director of the Underwater Ordnance Systems Group within the Bureau of Ordnance. Reich's responsibilities in the new position included oversight of the Torpedo Branch, the Mine Branch, the Fire Control Branch, and the Launch Branch. Shortly after becoming director of Underwater Ordnance Systems Group, Reich received a report concerning the defective performance of the Mark 14 torpedoes, including failures of the depth mechanism, erratic torpedo runs, and warheads failing to detonate during practice firings. For Reich, who had been a first-hand party to the Mark 14 problems during World War II, it was a familiar situation that was hard to understand given the fixes instituted during the latter part of the war. After discussing the issue with Capt. James M. Robinson, head of the Underwater Ordnance Planning Division of the Bureau of Ordnance, the two men agreed on the need to take quick action. They assembled a group of technical experts who were sent to Hawaii to investigate the problem.

After establishing themselves in the torpedo shop on the submarine base in Pearl Harbor, the technical experts began conducting an examination of how the torpedoes were being prepared and fired during practice. In order to establish exactly what was happening, Lt. Cdr. Henry Gottfried, an ordnance engineer assigned to the Bureau of Ordnance's production branch, set up a test program involving fifteen Mark 14 warshots. These were fired against the cliffs on the unoccupied Hawaiian island of Kahoolawe, duplicating tests that had been conducted against these same cliffs in August 1943. Just as had happened during those earlier experiments, the first few shots produced a number of duds. When these dud torpedoes were disassembled after recovery, the inspectors found that the springs used to control the depth mechanism were out of spec. When the manufacturing history of the springs was traced, it was discovered that after the production contract had been terminated, the manufacturer shipped what was left of their stock to the Navy without bothering to conduct any quality control checks. This material found its way into the stores. This explained the faulty depth mechanism. The torpedo specialists sent to Hawaii also found faulty material in some of the exploder mechanisms as well. Once the defective parts discovered by the specialists were replaced, all subsequent torpedo firings proved to be successful.[12]

In the fall of 1954, Reich got a call from Captain Robinson, who asked Reich to come in to discuss a special project the CNO wanted the Bureau of Ordnance to pursue. The OpNav planning group had determined that the

Chinese Nationalists in Taiwan needed a mining capability to deter the Chinese Communists on the mainland from invading. To supply the mines, the Bureau of Ordnance would develop a moored contact mine fitted with a booby-trap triggering device that had been used in the Far East during the latter stages of World War II. Lt. Cdr. Richard B. Plank, an ordnance officer in the Bureau of Ordnance who had been involved in the development of this device, was chosen as the project officer. The assignment given Commander Plank was to come up with a simple, highly reliable moored mine that could be manufactured inexpensively. The program was christened Project Timber.[13]

Plank's first task was to produce a prototype of the triggering device, which Reich claimed was "rather crude," but it was very sensitive and was thought to be very reliable. The Bureau of Ordnance issued a small contract to the Hamilton Watch Company to manufacture the device. The next issue faced by Plank was to come up with suitable material for the case. The most economical approach was to use a commercially available 55-gallon steel drum. Plank made a deal with one of the large drum manufacturers to make small changes in the top and bottom covers so that the drums could be used as mine cases. The CNO planners, who were to take care of the project's finances, had established a target price of $100 per mine. An ordinary 55-gallon steel drum could be obtained for around $25 at the time. Adding the special changes required to make it suitable for the Bureau of Ordnance's requirements added another $10 dollars to the cost, bringing the total procurement cost for the case to around $35. The drum would be filled with explosives and would be detonated by the sensitive triggering device that was about two inches in diameter and twenty-four to thirty inches long. Concrete blocks would be used as anchors.[14]

It took around ninety days to work out the design, the work on which started sometime in September. By February 1954, enough mines had been produced to conduct a test of their immersion in sea water. Thirty test mines were planted in the waters of Key West, moored in one hundred feet of water, floating six to ten feet below the surface. The mines were loaded with inert main charges but retained the one-pound booster charges that would go off when the mine was struck. The explosive force of the booster charge was not strong enough to damage the net-layer *Yazoo* (AN 92) that was chosen as the test ship, but it would readily indicate whether the mine worked. The mines were left in place for ninety days before any testing was conducted to make sure there were not any problems related to immersion in sea water.

The National Security Industrial Association (NSIA‡), which happened to be chaired by the chief of the Bureau of Ordnance, was scheduled to have its meeting in Key West around the time that testing was scheduled to begin. This enabled the committee, which joined the *Yazoo*, to observe the tests firsthand. As the mines were located, the *Yazoo* would be sent over to detonate the mine by coming into contact with the firing mechanism and setting off the booster charge. Because the ballast tanks of the *Yazoo* were empty, the sound of the booster charges going off was so loud the committee members thought the ship was going to blow up.[15]

With the design of the Timber Mine proven, the Bureau of Ordnance procured enough material to ship ten thousand drums, chains, and firing mechanisms to Taiwan. Commander Plank and a small group of assistants were sent to Taiwan. After arriving, they set up a facility for assembling the mines and loading them with explosives. Plank's team also advised the National Chinese Navy with advice on how and where to establish mine fields surrounding the island.

As Reich noted, the Timber Mine "was the soul of simplicity" and "a very effective and cheap weapon." He considered these characteristics "somehow abhorrent to the Navy," because the mine was a dangerous weapon that did not have the built-in safety features that would have been incorporated if it had been built to Bureau of Ordnance safety regulations. Nevertheless, he considered the Timber Mine one of the great achievements of his tour in the Bureau of Ordnance. He had been able to produce an inexpensive weapon in a very timely fashion by securing the cooperation of the Hamilton Watch Company, the Naval Gun Factory, and the drum manufacturer.[16]

Reich was not the only one pleased with the project's outcome. The CNO was also pleased, so much so that he wrote a letter to the chief of the Bureau of Ordnance, Rear Adm. Frederick S. Withington, expressing "his appreciation to the naval and civilian personnel in the Bureau of Ordnance whose imagination and perseverance contributed to the successful completion of the project." Admiral Withington forwarded the letter to the commanding officer of the Naval Ordnance Laboratory, Capt. John T. Hayward. While Reich's name

---

‡ Established in 1944 as the National Security Industrial Association, the NSIA was a nonprofit and nonpolitical organization. It was established to foster a close working relationship and effective two-way communication between the major defense agencies of government and the industry supporting it.

was not specifically mentioned, his role in making Project Timber a success was undoubtedly noted by Admiral Withington and Captain Hayward, both of whom would later play important roles in advancing Reich's career.[17]

In early August 1955, Reich was detached from the Bureau of Ordnance and reported as student to the Industrial College of the Armed Forces at Fort McNair in Washington, D.C.

CHAPTER 9

FROM STUDENT TO BUDGET OFFICER, WITH SEA COMMANDS BETWEEN

THE INDUSTRIAL COLLEGE OF THE ARMED FORCES, which was established on April 11, 1946, evolved from the Industrial College of the Army, which had been founded in 1924 to educate officers for the procurement division of the War Department. Its reconstitution in September 1948, established it as an educational institution under the Joint Chiefs of Staff. The reconstituted Industrial College of the Armed Forces was placed on the same level as the National War College. The two institutions differed in that the National War College dealt with international relations and national military security, whereas the Industrial College emphasized the economic and industrial aspects of national security. The mission of the Industrial College, as defined by its charter approved by the Secretary of Defense, is as follows:

> To prepare selected officers of the Armed Service for important command, staff and planning assignments in the National Military Establishment and to prepare elected civilians for important industrial mobilization planning assignments in any government agency.[1]

Reich, whose shore duties had heretofore been largely concerned with weapons (torpedoes) and industrial mobilizations (reactivation of mothballed ships), was pleased to be selected for the school. Assignment to the Industrial College of the Armed Forces also boded well for Reich's chances at further promotion. Initially, Reich thought that the assignment to the college was going to be a sabbatical following his very intensive duties in the Bureau of Ordnance. He felt that it would be an opportunity to sit back and study a little bit more on aspects of industrial support for the military. It turned out to be a great deal harder than he had expected, with lots of assigned readings and numerous reports and papers to prepare. "It was not an easy course," stated Reich. The curriculum was "very demanding."[2]

By the time Reich entered the college in August 1955, he had developed some serious concerns—based on his experience—about how new weapons were produced and about the maintenance practices needed to keep them reliable once they were in service. He had come to believe that those in charge of new weapon development were not paying enough attention to the importance of manufacturing technology or the need to make research and development funds available to improve producibility. Reich was also concerned about the lack of attention to issues surrounding the modification, maintenance, and reliability of systems in service.

Given this thinking, it is not surprising that Reich chose to write his thesis on the transition from weapons development to weapons production. In addition to understanding the technical side of weapons, Reich stressed the need to have a detailed knowledge of how financial resources are made available by the budget system and the interplays within the executive branch of government and the Department of Defense, the Bureau of the Budget, and Congress. A good knowledge of the operations side of the military was also essential. "A program manager," Reich concluded, "should have a span of understanding of that whole universe."[3]

In late February or early March 1956, when Reich was still a student with several months of school remaining, he began calling the detailing officer in the Bureau of Personnel responsible for officer assignments to lobby for a favorable appointment for his next tour of duty. Because he had previously served as acting division commander of submarines, Reich felt that he was eligible to be considered for a submarine squadron command. The submarine detail officer, however, was of the opinion that because Reich had not had a full tour of duty as division commander, his eligibility for a squadron command was questionable. Then one day in March, much to Reich's surprise, he received a phone call from Assistant Chief of Naval Operations for Undersea Warfare Rear Adm. Frederick B. Warder. Warder, a highly decorated veteran World War II submarine commander, was undoubtedly familiar with Reich's combat record. Why Warder decided to call Reich remains unknown, but during the conversation he asked whether Reich knew where he was going after he completed the course at the Industrial College. Reich explained that he "was very much in doubt and also very unhappy because of the overtures [he] had made toward the submarine detail people in BuPers [Bureau of Personnel] hadn't been well received." Warder told Reich he would look into it. Two weeks later Reich received word from the submarine detail desk that they were cutting orders for him to take Submarine Squadron 8, based in New London, Connecticut.[4]

Reich left the Industrial College in June 1956 and arrived in New London in early July to take over the squadron, which was part of the Submarine Force, U.S. Atlantic Fleet. Reich was not afraid to take on additional duties if it would be beneficial to the Navy, and he knew how to work with other commands in order to achieve his objectives. Although Submarine Squadron 8's primary mission was to support the Atlantic Fleet, Reich realized that the Submarine School's base squadron could not handle the workload of underway submarine training, and he offered to take on the PCO training duty. He worked out the details of Squadron 8's PCO training with the base squadron, the Submarine School, and the staff of the Submarine Force, Atlantic Fleet. He believed that his squadron could benefit almost as much as the students if the advanced training were conducted in the Bermuda area as opposed to within the area off New London. After a few visits to the staff of Destroyer Force, Atlantic Fleet (DesLant), at Newport, Rhode Island, Reich managed to work out a mutual service agreement whereby DesLant would assign a division of destroyer escorts to work with Squadron 8 and to participate with the PCO training in the Bermuda operating areas.[5]

The arrangement worked out well. Reich would depart New London on Monday morning with five or six of his submarines, accompanied by the submarine rescue ship *Tringa* (ASR 16), and would head for the operating area south of the Gulf Stream that provided excellent winter weather conditions. Once in the operating area, Reich's force would be joined by five or six destroyer escorts. The *Tringa* would be loaded with exercise torpedoes and would function as a command ship and tender. Reich set up an operational control point in *Tringa* and assigned an operations officer on board. Reich would go from submarine to submarine and observe both the individual student approaches and the general conduct of operations. This provided intensive training in terms of the number of target approaches and the number of exercise torpedoes fired. The spent torpedoes were skillfully recovered by the *Tringa*, which performed this duty without difficulty. The destroyers served as target ships, and they also had the opportunity to practice their antisubmarine warfare skills.[6]

The trip down to the operating area took two days. Operations would begin on Wednesday and run through Friday afternoon. At which point they would steam into Bermuda and moor at the Royal Navy base so the crews could be given liberty on Friday night, Saturday night, and Sunday. Reich's force would steam out of Bermuda early Monday morning and be back in an operating area within a half hour of leaving Bermuda. After four days of exercises, the force would head back to New London and be back in port by the following Friday.

Reich enjoyed his tour as Squadron 8's CO, but command of a submarine squadron at that time was not considered a major command. In order to be eligible for a major command, Reich needed to complete a successful tour in command of a deep-draft ship.* For Reich, this was to be the fleet oiler *Aucilla* (AO 56), a *Cimarron*-class tanker constructed during World War II that drew 32 feet of water.

In August 1957, Reich was relieved from command of Submarine Squadron 8 and received orders to report to the *Aucilla* in Baltimore, where she was undergoing repairs in a civilian yard. After the repairs were completed in mid-September, Reich took the oiler to the Guantanamo Bay operating area for refresher training. Reich, as usual, wanted to accomplish more than was typically required, and he arranged for *Aucilla* to serve as a target ship for some of the submarines undergoing training there. It provided the submarines with the opportunity to use a deep-draft ship as a target, and it gave *Aucilla*'s crew experience in deck handling, seamanship, and maneuvering when recovering the spent practice torpedoes. The exercises were a novelty for many of the officers and crew since few of them had ever seen a torpedo heading for a ship. The ship returned to Norfolk after completing refresher training and departed again on October 28, 1957, this time heading to Barcelona, Spain. Barcelona was to be her home port while she was assigned to the Sea Logistics Europe Command, with duty supporting the operations of the Sixth Fleet in the Mediterranean.[7]

The 553-foot long, 23,000-ton *Aucilla* was a so-called white tanker because she carried only aviation products. Her main cargo tanks were filled with Avgas and jet fuel. The ship's mission was to keep the regular fleet oilers filled with aviation fuel. The Navy had a contract with an Italian-owned refinery in Augusta Bay, Sicily, to offload a certain amount of jet fuel or Avgas from the

* A deep-draft ship requires lots of water under its keel, which means special skills are needed to navigate it in shallow waters such as harbors and channels.

refinery's storage tanks every month. After filling her tanks at the refinery, *Aucilla* would rejoin the Sixth Fleet to resupply the Navy's regular tankers. The regular tankers would discharge half, or more than half, of the Avgas or jet fuel they carried every time they came alongside one of the carriers for underway replenishment. They had enough heavy oil for the numerous refueling evolutions required to top off the ships of the Sixth Fleet, but they had only a limited supply of aviation fuel. Using *Aucilla* as a roving gas station allowed the oilers to remain on station longer. *Aucilla* shared this duty with the fleet oiler *Marias* (AO 57), and one or the other of the oilers arrived at the refinery every fourth or fifth week, if not sooner.[8]

During his tour as captain of the *Aucilla*, Reich had to deal with a number of personnel problems that are inherent in a fairly large crew. His executive officer, whom Reich characterized as "a very personable fellow, very articulate and very proud of the ship and the ship's company . . . was a public relations man . . . who was very heavy on PR" but very short on leadership responsibilities. Reich made sure that he was relieved before the end of his tour. Much to his chagrin, Reich also discovered that half of the chief petty officers on the ship were more interested in drinking coffee in the chief's quarters than they were in getting off their duffs and "turning to." He was particularly annoyed with the chief electrician, who "didn't know much about gyros," and who did not pay much attention to the problems they were having with the ship's gyro compass. The arrival of Chief Warrant Officer Chester T. "Chet" Sablowski solved Reich's petty officer problem. Sablowski had been to sea a lot and "was hard as nails." He was not on board more than two weeks before he had the chief petty officers straightened out. With Sablowski's help, Reich saw to it that the chief electrician was disrated and discharged from the ship.[9]

In the early part of summer 1958, Reich, who was still in command of *Aucilla*, received a letter from Rear Adm. Frederick B. Warder, the commander of Submarine Force, Atlantic Fleet. In his letter, Admiral Warder stated that his current chief of staff had just received orders to command a cruiser, and that he [Warder] would be very happy to have Reich as his replacement. Reich responded with a letter of his own saying that he would be delighted and was grateful for Warder's vote of confidence. Reich was very happy for the next few weeks, believing that he would soon be going back into the submarine community as chief of staff for ComSubLant and working for a man he considered to be "an outstanding officer."

Reich's euphoria did not last very long. In the latter part of August 1958, he received a letter from Rear Adm. John T. Hayward, the assistant chief of naval operations for research and development. In his letter, Admiral Hayward stated that Chief of Naval Operations Adm. Arleigh A. Burke was about to create a new organization to manage the development of new weapons that would be headed by a deputy chief of naval operations for development holding the rank of vice admiral. Hayward was going to be promoted and was going to be the first deputy chief of development, and he wanted Reich to assist him in organizing and setting up this organization.[10]

Hayward had known Reich from his days in command of the Naval Ordnance Laboratory in White Oak, Maryland, when Reich was serving in the Bureau of Ordnance. Reich should have been pleased that the admiral thought enough of his capabilities to invite him on board the new organization. Working for a vice admiral was not anything to sneeze at. But Reich had argued and feuded with Hayward regularly over various matters having to do with underwater ordnance and fire control. Reich, in a long letter said, "No thanks," telling Hayward that he had already committed himself to Admiral Warder's staff job.[11]

Shortly thereafter, Reich received a note back from Hayward saying that he had gotten in touch with Admiral Warder who told him that since his present chief of staff's orders had been delayed for at least six months, Reich was once again available. If Admiral Hayward wanted Reich, Warder had maintained, it would be alright with him. And that was that. In October 1958, Reich was relieved of command of *Aucilla* and received orders to proceed to the Office of the Chief of Naval Operations and report to the deputy chief of naval operations.

Admiral Hayward was a very smart guy who had vast experience with the development of naval weapons. He had participated in the Manhattan Project that developed the atomic bomb, had traveled to Hiroshima and Nagasaki as part of a team investigating atomic bomb damage, and had led the effort to photograph the nuclear explosion at Bikini Atoll during Operation Crossroads; he was also the first naval aviator to command the Naval Ordnance Laboratory and, at forty-six years of age, was about to receive his third star after serving as a rear admiral for only two years. While Hayward had become acquainted with Reich through their contact during Hayward's tour in command at the Naval Ordnance Laboratory, it is inconceivable that he would not have looked more closely at Reich's record and qualifications before offering Reich the opportunity to take part in the establishment of a new, high-level branch of OpNav. Reich's Industrial College thesis, his two tours in the Bureau of Ordnance, and his ability to get things done made him the ideal candidate.[12]

The need to establish the CNO's Office for Development came about as part of the Department of Defense's recognition that there was a great deal of redundancy in the manner in which research and development, and testing and evaluation were being handled within the armed services at the time. The solution, according to the senior leaders at the Pentagon, was to create a new program within the Department of Defense to coordinate and consolidate research, development, testing, and evaluation within a single organization. Admiral Hayward's job was to do the same within the Navy by establishing a single Navy budget for research, development, testing, and evaluation (RDT&E), that is, to centralize control over all RDT&E funding within the Navy. Reich's job was to take charge of the Program Management and Budget Office within the CNO's Office for Development.[13]

Reich now faced the daunting task of trying to pull together the separated funding requests made by each of the bureaus and, using this data, allocate funding among the budget categories for research, development, testing, and evaluation. None of the bureau chiefs, of course, relished the idea of having their appropriation budgets for these items taken out from under them. Reich was starting from scratch, since there had never been a single RDT&E budget for the Navy. To accomplish this task, Reich had eighteen or twenty people to help him, including three or four junior captains, two or three civilian accountants, and a number of commanders. Reich's team had to organize all the data into categories laid down by the Navy's Comptroller's Office.[14]

The budget inputs had to go through a so-called sponsor in one of OpNav's branches. Reich's office would put all the branch budget requests together and submit the package to the Comptroller's Office. That office would then establish an overall limit on the amount of money that could be requested by the Navy for any particular item in the fiscal year. To stay within the budget limit established by the Comptroller's Office, decisions had to be made as to which items needed to be kept in the budget and which could be removed. Such decisions were made during hearings conducted by the CNO Advisory Board (CAB). The director of the Navy budget and his boss, the deputy comptroller, would set up a big blackboard in the CNO conference room for a budget meeting chaired by the vice CNO. Together, the director of the budget, the deputy comptroller, the

CNO, and his deputies, each with their supporting assistants sitting behind them, would go through each appropriation, arguing which should stay in and which should be removed.[15]

On a number of occasions during the process of establishing the RDT&E budget, Reich would be called into Admiral Hayward's office to provide the latest numbers on the budget. Reich would give him the latest figures, and no matter what they were, Hayward would tell him to find a way to reduce it by $100 million. And he wanted it done by 1700. Reich would go back to his office and round up all the lead people who headed up the various warfare areas in the office. He would tell them, "Fellows, we've got to take $100,000,000 out of the budget that we racked up last Friday. Now here is the procedure for today. We're going to take $200,000,000 out of it—right in this office. . . . I am arbitrarily going to assess each one of you in your areas a number."[16] Reich did not intend to take $200 million out of the budget, but he wanted to illuminate some alternatives for Hayward's benefit. Reich recalled,

> I'd tell them that I didn't want any arguments and wasn't asking them if they thought my arbitrary allocations were right, wrong or different; I was just giving them numbers. Then I'd tell them to go out and lock their doors and the only thing they could use was the telephone because I didn't want anybody in or out of the office as we were going to do this within our own shop. If they needed to get information from somebody, then use the telephone, but nobody gets out of here until we finish this exercise.
>
> I'd tell them to go study their program areas and see how they were going to do it and when they got their solutions to come into my office and we'd discuss it. Then we would put it on the blackboard as to what particular project [it was] and how you could get $200,000,000 out of this one, or if you could cancel that one and what you're going to do with the other one, and so on.[17]

Nobody was happy with that kind of advice, but, as Reich said, "Off they'd go."

When the time came for Hayward to review the finalized budget, he would come over to Reich's office and close the door. He would sit at a table as Reich went over the details laid out on a large blackboard. Various senior members of Reich's staff were called in as needed to defend their pieces of the pie. After much give and take, Hayward would go to the blackboard and either check or change various things. When they were all done, Hayward would look it over. "Well, okay," he'd say, "that's it—I need this all on paper for a Board Meeting tomorrow morning at 9:00 a.m." It was usually around 1630 when they finished up and Hayward left the office. Reich and his staff would stay until 2000 or 2100 putting it all together.[18]

Reich's biggest headache was having to deal with the likes of Deputy Chief of Naval Operations (Air) Vice Adm. Robert B. Pirie, and the other strong-minded deputy chiefs of naval operations. During one vitriolic encounter with Admiral Pirie at a CAB hearing, Reich had to be rescued by Adm. James S. Russell, the vice CNO, who told Pirie, "Now Bob, you've had your say and I understand and I think Captain Reich understands what you've said, but on the other hand, you've heard his rational and, Bob, I think that's the way it's going to be."[19]

In August 1960, a little over two years since he took over as head of the Program Management and Budget Office for RDT&E, Reich, who was now eligible for a major command, received orders to command the guided missile cruiser *Canberra* (CAG 2) thanks to positive endorsements and support from both Vice Admiral Hayward and Admiral Russell.

CHAPTER 10

COMMANDING A GUIDED MISSILE CRUISER

THE *CANBERRA* (CAG 2) WAS AN IMPORTANT COMMAND. The World War II heavy cruiser had been converted into a guided missile ship by the New York Shipbuilding Corporation in Camden, New Jersey. She was only the second such ship in the Navy when she was recommissioned on June 15, 1956. During the conversion, her aft 8-inch turret was removed and replaced with two Terrier dual missile launchers, each with its own automatic loading and handling system. Two missile magazines dubbed "Coke machines," because of their layout, were added below the handling rooms. The ship's two funnels were trunked together, and the forward superstructure was simplified. A large, lattice foremast carrying an SPS-8 height-finding radar was added along with a new pole mainmast for the CXRX hemispherical search radar. Fire control was provided by two SPQ-5 monopulse C-band beam-rider guidance radars.[1]

The Terrier was a supersonic, beam-riding, surface-to-air anti-aircraft missile that entered service in 1956. The SAM-N-7 BT-3* variant that Reich indicates was on the *Canberra* weighed 3,000 pounds, was propelled by a solid-fuel rocket booster and a solid-fuel sustainer, had a range of 15 nautical miles, flew at Mach 3, and carried a 218-pound controlled fragmentation warhead.[2]

In August 1960, when Reich received orders to assume command of *Canberra*, he was unaware of the ongoing hardware and reliability problems the Navy was experiencing with the surface-to-air guided missile systems that had recently been added to the fleet. Early exercises on the *Boston* (CAG 1), *Canberra* (CAG 2), and *Galveston* (CLG 3) revealed operational problems. In tight formations, falling boosters endangered other ships; sorting friendly aircraft from enemy planes was frequently a problem; the missiles were vulnerable to electronic countermeasures, and their kill rates were abysmally low. Despite

* Redesignated the RIM-2C in 1962.

these problems, there were some in the Navy's hierarchy who did not want to acknowledge or advertise the deficiencies in their commands. This was especially true for Rear Adm. John McNay Taylor, commander of the Atlantic Fleet cruiser force. On March 2, 1960, Taylor sailed from Norfolk on board his flagship *Canberra* to start a goodwill cruise around the globe so that *Canberra*, one of the first guided missile cruisers in the Navy, could engage in anti-aircraft warfare (AAW) demonstrations and tactical exercises with the Seventh Fleet in the Pacific and the Sixth Fleet in the Mediterranean. Taylor, according to Reich, "was concerned [that] Navy missilry and advanced AAW was out in the forefront, and he was there to demonstrate that success to all." Taylor's views were reflected in the numerous dispatches filed during *Canberra*'s cruise that touted the outstanding performance of the ship's latest procedures and new tactical concepts. "There was an aura of excellence and great success, with Admiral Taylor's name very much tied to all of this accomplishment and advance." In retrospect, it appears that Taylor, whose name would soon appear before the Selection Board for vice admiral, wanted his current command to be shown in the best light.[3]

Reich arrived in Norfolk several days before *Canberra*'s return to port and reported (in Admiral Taylor's absence) to Taylor's chief of staff, Capt. Phillip F. Hauck. Hauck, an Annapolis classmate, explained that *Canberra* was not due in port for another week. He briefed Reich on the command's organization and suggested that as the ship would not be in port for a few days Reich should just relax and let the staff continue with their briefings.[4]

The *Canberra* arrived in Norfolk on October 24, 1960. As one would expect after such a long deployment, there was a large crowd of family and friends waiting to greet the ship when she returned to her home port. Reich did not go near the ship that first day "because they had their hands full." When he arrived on board the next day, he was shown into Capt. Walter H. Baumberger's cabin. Admiral Taylor, Baumberger explained, was due to be relieved in two or three days, and on his relief, he was going to be elevated to vice admiral and assigned command of the amphibious force of the Atlantic Fleet. Baumberger advised Reich that all his time during the next few days would be taken up in making arrangements for the elaborate change in command ceremony that was going to take place and suggested that Reich take the next few days to review the ship's records. Reich thought this was a good idea, bundled up all the ship's reports and organization books and made himself scarce for four of five days.[5]

Captain Baumberger's reports were very thorough and lengthy. From his previous duty, which entailed the review of the budget for the Terrier, Tartar, and Talos systems (the 3Ts), Reich knew that these programs were not performing well. Now, as he waded through Baumberger's material, it became evident that as far as the performance of *Canberra*'s missile battery was concerned, things were not going very well—neither in terms of system availability nor in terms of the actual performance of the missiles themselves, which displayed a disturbing inability to intercept their targets. Reich knew that Admiral Taylor considered Navy missilry and advanced AAW to be at the forefront of the Navy's public persona, and "he was there to explain and to demonstrate that success to all and sundry; but when one took a hard look at what was written on paper—as far as reflecting the actual performance—there was a lot to be desired."[6]

After assuming command of *Canberra* on November 11, 1960, Captain Reich asked Rear Adm. Robert W. Cavanagh, the newly installed commander of the Atlantic Fleet cruiser force, if he could take the ship to sea in December. Reich was anxious to get his ship to sea before the Christmas holidays to "shake it down a little bit and to see [how] the ship perform[ed] at sea." He also asked if he could conduct some missile firings off the Virginia Capes. Admiral Cavanagh agreed with Reich's request and allowed him to take the *Canberra* out during the last week in November.[7]

Neither Reich's weapons officer nor his missile officer shared the superlatives that had been used to describe *Canberra*'s performance during her around-the-world cruise. They, too, had their doubts about the ship's missile system and greeted Reich's plan to conduct more firings with approval. As Reich later explained, "It wasn't a dress rehearsal but rather it was like a postmortem we were going to do." So, he took *Canberra* to sea in late November and conducted test firings of the ship's Terrier missiles while steaming off the Virginia Capes. Of the seven or eight missiles fired, six were aborted or failed to perform satisfactorily. The performance of the Terrier missiles was almost identical to what had occurred during the demonstration exercises conducted during Admiral Taylor's good-will cruise.[8]

After returning to port, Reich arranged to see Rear Adm. Charles K. Bergin, commander of the Operational Tests and Evaluation Force. Reich had served under him during his tour of duty as head of the Torpedo Branch in the Research and Development Division of the Bureau of Ordnance, and

Reich wanted advice, counsel, and help.[†] Bergin agreed to help Reich obtain data on all the missile firings that *Canberra* had ever undertaken. This data were obtained from the Naval Ordnance Laboratory (NOL) in Corona, California, which collected and analyzed records of all the missile firings that took place in the fleet. During his Christmas vacation, while the ship was standing down, Reich studied the data gathered by NOL's analysis group on *Canberra*'s missile firings. After reviewing data from the 120 Terrier missiles *Canberra* had fired since her commissioning, Reich concluded that the beam-riding Terrier SAM-N-7 BT-3, which was the first Terrier to be deployed, had never performed in a reliable, successful way.[9]

For Reich, the problems relating to Terrier's reliability must have reminded him of his experience in Turkey with the reported failures of the Mark 14 torpedoes. The solution back then was to conduct a series of highly supervised firing tests under controlled conditions. Reich was about to suggest a similar approach for Terrier.

After the Christmas holiday, Reich traveled to Washington, D.C., to see his former boss, Deputy Chief of Naval Operations for Development Vice Adm. John T. Hayward, to discuss the missile situation. After talking the matter over, they came to the realization that they were not going to convince anybody of the problem based on Reich's analysis alone. But they could set up a test firing under controlled conditions during Operation Springboard, the annual training exercise for ships that deployed from the East Coast to the Puerto Rican area to conduct gunnery, ASW, and AAW exercises. The Bureau of Naval Weapons[‡] (BuWeps), however, was opposed to the idea. BuWeps felt it was a needless expenditure of expensive weapons, was outside the normal fleet training exercise, and was a gross misapplication of inventory assets. Reich thought that part of the problem resulted from the exultation that existed within the higher echelons of the Navy, who believed that it was a great navy, a wonderful navy, and that anybody who suggested that it might not be so in certain areas was not well respected. With Admiral Hayward's support, and after considerable

† When Reich had served under him, Bergin held the rank of captain and was assistant chief of the Bureau of Ordnance for research from the summer of 1953 until December 1955.

‡ The Bureau of Naval Weapons was created on July 1, 1960, by an act of Congress. It consolidated the activities of the Bureau of Ordnance and the Bureau of Aeronautics into the new organization.

bickering within both OpNav and the Bureau of Naval Weapons, Reich received permission to set up a missile firing test during the upcoming Springboard exercises. *Canberra* would be given a drone unit that would be embarked so they would have their own target. She would be authorized to expend twenty-five missiles and would be required to prepare a technical schedule that would be reviewed by the Applied Physics Laboratory (APL), the developers of Terrier who were now BuWeps technical advisers for the system.[10]

Reich sailed the *Canberra* to Roosevelt Roads Naval Station, Ceiba, Puerto Rico, in January 1961. From there he conducted missile firing tests while operating seventy-five miles southeast of Puerto Rico. He had complete control of the drone unit but had to clear it with Range Control whenever the ship was firing. The tests went off as planned, although they ran into quite a few problems as the exercises progressed. Special observers from APL, the Sperry-Rand Corporation (the radar supplier), and BuWeps were on board to witness the tests. They had some success, but it was spotty. Additional instrumentation had been added to the Terrier system, and each missile firing was analyzed by telemetry experts who had been embarked on the ship so that the data collected could be quickly evaluated. After the exercise was completed, Reich sent a post-firing report to the commander-in-chief of the Atlantic Fleet. Information copies were also forwarded to the BuWeps and the CNO. The tests, he wrote, confirmed his previous observations "that there were some very serious defects in the [Terrier BT] system, whether they be fire control, radar, or missile performance."[11]

The people in BuWeps, who had not been overly enthusiastic about the tests to begin with, were quite skeptical of the results Reich reported. The tests, according to the folks in BuWeps, had not defined or demonstrated anything new. Yes, there were defects, and yes, this kind of performance could be expected, but they were already planning modification of the fire control radars, the weapons direction equipment, and the missile. These problems would all be resolved when *Canberra* underwent her normal overhaul at the Norfolk Naval Shipyard in Portsmouth, Virginia, scheduled to begin in May. The official reaction to the evidence submitted by Reich was that the folks who really knew the technology and who were responsible for the technical cognizance of the Terrier were those in BuWeps, and they had the situation well in hand.[12]

Shortly after the ship's arrival in Norfolk, Reich and his missile officer met with Capt. William R. Crenshaw, the yard's ordnance superintendent, to discuss the changes that were needed to improve the ship's missile system. Reich, aware of the basic defects in his ship's missile battery, was very anxious to have the modifications recommended by BuWeps installed. It quickly became apparent to Reich that the technical groups within BuWeps were uncertain about the fixes needed and that there was a lack of responsibility toward in-service missile systems within the bureau. Neither BuWeps nor the Bureau of Ships, in Reich's view, considered it a high priority. It seemed to Reich that they were leaning very heavily on the contractors to supply the technical expertise needed to make the fixes.[13]

Crenshaw was aware of the problems they were having in trying to get the work done. Although he had made strong requests from the Bureau of Ships and the Bureau of Weapons for technical guidance and assistance on the work to be done—both in writing and by telephone—their response had been slow and incomplete. Nothing was heard from these bureaus for about forty-five days. Then, suddenly, the yard was swamped with a rash of paper instructions detailing the changes that needed to be made. Some new equipment was sent to the yard along with four or five engineers from Sperry-Rand to assist with the changes. By then, the midpoint of the overhaul had passed and there was a frantic rush to complete the missile work. As Reich and those members of the crew who were familiar with the missile system observed the work being done, the quality of the workmanship left them with serious doubts as to whether the system had been adequately tested and fixed.[14]

Before leaving the yard, Reich discussed the overhaul with its commander, Rear Adm. William E. Howard Jr. Reich told Admiral Howard that he thought the material overhaul of the ship had been excellent and the dealings with the ship's company had been good, but that his ship was leaving the yard with its main battery in an unsatisfactory condition. Reich said that while he understood that the yard was not responsible, he intended to make a formal report to the commander-in-chief of the Atlantic Fleet. When *Canberra* left the yard at the beginning of September, the problem with the missile systems had not been corrected, and none of the missile systems was operational. Many of the civilian contractors had to stay with the ship in order to complete the modifications needed to make these missile systems fully operational.

Back at the midpoint of the overhaul, Reich had contacted Admiral Hayward to point out the problems with the overhaul and his concerns regarding the effectiveness of any of the changes recommended by BuWeps. Reich conveyed to

Admiral Hayward his belief "that the bureau people and the contractors were not really concerned with in-service weapons; they were more concerned with things that were going to come out in the future, and the problems of in-service systems [were] supposed to be somebody else's [the contractors'] responsibility." He suggested that they ought to repeat the fleet tests that had been conducted the previous spring.[15]

Admiral Hayward and Admiral Bergin were again largely responsible for overcoming the ongoing opposition from the Office of the Chief of Naval Operations (OpNav) and BuWeps, which mirrored their previous objections to Operation Springboard. By the time *Canberra* was ready to leave the yard, Reich had received orders to conduct the usual post-overhaul training while proceeding to the Guantanamo Naval Base, in Cuba. From there, he was to proceed to Roosevelt Roads and the Atlantic Missile Range to conduct a fleet missile test program like that conducted during Operation Springboard.

During the cruise to Guantanamo and on to Roosevelt Roads and the missile range, *Canberra*'s crew, with the help of the contractors from Sperry-Rand Corporation, Vitro Corporation, and APL, readied the ship's missiles, radar, and fire control system. They arrived in mid or late October and began the test program that involved the twenty-five Terrier missiles allocated for this purpose. "It was almost a replica of what we had done during Springboard," Reich declared years later. "Again, the overall performance was the same, if no worse than the performance in Springboard."[16]

When the *Canberra* arrived back in Norfolk on Sunday morning, November 25, 1961, there was a staff officer from the Cruiser Force, Atlantic Fleet, waiting alongside the pier with instructions for Reich to report to Rear Adm. Michael F. D. Flaherty, who had just relieved Admiral Cavanagh as cruiser commander. Reich and his missile officers and weapons system officers were also instructed to go to BuWeps on Monday morning. Reich enjoyed a cordial meeting with the admiral at his quarters later that day and provided a detailed account of his experience with *Canberra*'s missile systems, telling the admiral, "These systems were really sick and that it was a charade to continue the notion that band-aid [*sic*] fixes would do."[17]

That Monday, Reich along with his executive officer, Cdr. Merrill Sappington, arrived at the BuWeps offices for their 1000 scheduled meeting. Upon arrival they were shown into a large conference room. When the meeting convened, there were thirty-five to forty people in attendance representing various groups in the Bureau of Ships, BuWeps, and the Division of Fleet Maintenance—half military and half civilian—all there to discuss Reich's numerous missives on the poor performance of the Terrier missile system.[18]

During the discussions that took place at this meeting, Reich found that there were still some people in the Navy's bureaucracy who felt that he was just out to upset the apple cart and rock the boat. These folks believed that things were not that bad and that—if those in the fleet would just let the bureaus work—the problems would be corrected in due course. Comments such as these provoked Reich into a few polemics of his own.

In Reich's view, there were fundamental design problems with the 3Ts, and no amount of minor fixing was going to correct the situation. The missile systems were not up to par, and the Navy, especially its technical people, were not paying attention to the existing problems. Instead, they were more concerned with the new systems that were going to come into being five or ten years down the road. Reich's assessment of these problems had the potential to threaten the careers of some of the officers present. In their minds Reich's criticism suggested that the Navy and their own organizations, in particular, were not perfect. As Reich put it "everything we did wasn't apple pie. . . . They were being compelled [to fix the problems] because of these tests and because of the activities in *Canberra* and because of these messages—it was sticking out like a sore thumb on the Washington Navy scene, and it had to be coped with."[19]

When Reich left the meeting, it appeared to him that his involvement with the Terrier problems was over. He was moving on. He knew that he would soon be relieved from command of the *Canberra* with orders to report to the Bureau of Weapons. While *Canberra* was still undergoing overhaul in the Norfolk shipyard, Reich had been informed that he had been selected for flag rank by the Selection Board. The selection for flag rank meant that Reich was due to be relieved from the ship. As the *Canberra* was returning from the test firing program conducted at Roosevelt Roads, Reich received a communication from the Bureau of Personnel that said upon his arrival he would probably be assigned to relieve the chair of the Ship Characteristics Board in OpNav. Reich did not like the news.[20]

CHAPTER 11

ELI REICH, GUIDED MISSILE CZAR

WHEN REICH RECEIVED WORD that he was being assigned to the chairmanship of the Ship Characteristics Board, he was not very happy about it. He had been before the board many times when he was in the Bureau of Ordnance, and he was very familiar with its operation. He knew that as chair he would have to handle a lot of diverse groups, but as he later recalled, "You really didn't carry a significant clout if your signal number was fairly low on the totem pole." Although Reich had been selected for promotion to rear admiral, he was still technically a captain with a signal number* that placed him 683rd in seniority. He was a "fresh-caught selectee," in his words, someone who was not expected to rock the boat. Wherever the Navy wanted to send him would be OK.[1]

On the day that Reich arrived in Washington, D.C., to attend the conference in the Bureau of Naval Weapons (BuWeps) on the Terrier missile, he was notified that instead of the Ship Characteristics Board, he was being assigned the director of space and aeronautics within BuWeps, serving under the chief of the bureau, Rear Adm. Paul D. Stroop. Reich was ecstatic. He returned to *Canberra* in Norfolk after the conference and was relieved of command during the traditional change-in-command ceremony on board. He left the ship around the fourth of December 1961 and went back to his home in Chevy Chase, Maryland, feeling pretty good. He had had an interesting, but demanding, and at times a very exciting cruise on the *Canberra* and was looking forward to his new assignment.[2]

On December 7 or 8, 1961, Reich received an envelope by registered mail containing a copy of a letter from Admiral Flaherty, commander of Cruiser Force, Atlantic Fleet, addressed to the commander in chief of the U.S. Atlantic Fleet reprimanding Reich. In the letter, Flaherty chastised Reich for his

* A signal number is the naval officer's numerical order on the official seniority list.

impudent, if not insubordinate, airing of fleet and cruiser problems during the conduct of the missile tests. Flaherty assured the chief of the Atlantic Fleet that there had been a grotesque exaggeration of readiness conditions in the *Canberra* and the Terrier Missile System. As the reader will recall from the previous chapter, Reich had met with Admiral Flaherty after Reich's last cruise and had provided a detailed account of his experience with *Canberra's* missile systems.[3]

Fortunately, Reich had taken his personnel file with him when he left *Canberra*, and he therefore had copies of the reports on the three missile tests that had been conducted. Reich spent the next week composing a letter addressed to the deputy commander of the Atlantic Fleet, Vice Adm. William M. Beakley, explaining what had gone on during the cruise and the results of the missile tests. He mailed his response on December 13 or 14. After ten days had passed, he received a Christmas card from Admiral Beakley. Written inside was an acknowledgment that he had received Reich's letter, which had been placed in the Commander-in-Chief, Atlantic Fleet's (CincLantFlt) files, and assurances that there was nothing to worry about. The admiral hoped that Reich's family would enjoy a very happy holiday. And that was the last the Reich heard of the reprimand or of Admiral Flaherty.[4]

When Reich had arrived at BuWeps on December 15, 1961, Admiral Stroop was away, so Reich reported to his deputy, Rear Adm. Kleber S. "Chief" Masterson. "Eli," said Masterson, "there are a lot of places we could use you. I sure wanted to get you into the missile business," but we had to fill the spot in Space and Aeronautics. Masterson strongly suggested that Reich go see Rear Adm. Jack P. Monroe, the head man in OpNav Space and Aeronautics, as soon as Reich got settled in. "Don't get too far in your job without checking with Admiral Monroe," he said.[5]

A day or two later, Reich had gone over to the CNO's office and arranged a meeting with Admiral Monroe. One of the first things Monroe wanted to know was whether Reich had ever been downrange? Reich did not know what he was referring to until Monroe described the various installations throughout the Pacific that formed the Pacific Missile Range. Admiral Monroe made it very clear that he expected Reich to be on the next flight that went downrange. So, on January 2, 1962, Reich boarded a Constellation aircraft at Point Magu Naval Air Station for an inspection trip of the Pacific Missile Range that included stops at Johnson Island, Kwajalein, Eniwetok, and Roi-Namur.[6]

While Reich was away, conducting his inspection of the Pacific Missile Range, significant events that would have a momentous impact on Reich's future career in the Navy took place in Washington, D.C. Reich was evidently not the only person or entity growing increasingly concerned over the Navy's guided missile program. The program managers at the Applied Physics Laboratory (APL) that had designed the 3Ts were also growing concerned. They first became aware of the systemic problems within the Navy's surface-to-air missile program during the operational evaluation tests conducted on the guided missile ship *Dewey* (DLG 14). *Dewey*, which was launched at Bath Iron Works, in Bath, Maine, on November 30, 1958, was the first U.S. Navy vessel designed from the keel up as a Terrier missile ship. She was also the first to carry the tail-controlled BT version of the Terrier missile that was scheduled for installation in all succeeding Terrier guided missile ships.

When the APL representatives arrived on board the *Dewey* in the early months of 1960 to assist the Navy with their operational evaluation of the Terrier system, they discovered a myriad of problems. These appeared to be due to the failure to test the system as a whole. During the evaluation period, APL's engineers uncovered problems with the ship's tracking radar, the guidance radar, and the missile's capture beam. "You couldn't keep the radar on the air," noted one of APL's men, "because some component would fail. People would stay up day and night trying to get the radar to stay on the air long enough to fire a missile." Or the computer would go down; or the launcher; or the missile—which had 100 vacuum tubes and 1,000 resistors, all of which had to function under widely varying conditions of shock, humidity, temperature, and pressure—would malfunction. The most critical breakdown was the fire control radar, which the ship's crew could not maintain in working order. To remedy this situation, APL established a remedial program, working with the various equipment suppliers and APL's ship qualification assist teams to ensure that all the systems needed to guide and fire the missiles, as well as the missiles themselves, were in full operating condition before a guided missile ship deployed. APL also established a fleet failure reporting system. The problem was too large for APL to solve on its own, however. APL's leadership became so alarmed with the growing deficiencies in the Navy's surface-to-air missile program that APL's director Ralph Gibson wrote directly to Admiral Stroop outlining the magnitude of the problems facing APL and the Navy as it tried to rectify the problems associated with what was becoming known in some quarters as the "Terrible Ts."[7]

Milton Shaw, a mechanical and nuclear engineer who had been a key civilian official in the development of the Navy's nuclear program, had also been investigating the Navy's missile program for Assistant Secretary of the Navy (for Research and Development) James H. Wakelin Jr. Shaw had become familiar with the problems associated with the 3Ts during the fitting out of the nuclear-powered *Long Beach* (CGN 9) and *Bainbridge* (DLGN 25). He was working as Wakelin's technical assistant when he issued his findings at the end of December 1961 or in the early part of January 1962. In his report, Shaw concluded that none of guided missile systems on the twenty-eight missile ships in commission could be called operative, and of the six Talos cruisers in the fleet, only *Galveston* was truly operational. According to Shaw's report, 6 billion dollars' worth of the Navy's ships were equipped with defective missile systems.[8]

Shaw's report was passed on to Rear Admiral Stroop. Shaw's report, in concert with Gibson's letter, caused major concern for Stroop. The admiral did not think much of Shaw's "scurrilous document," but it was raising hell with the Navy's tactical missile systems and making lots of allegations that he thought were groundless. Nevertheless, after discussing the seriousness of the problem with both the CNO and the Secretary of the Navy, Stroop decided to create a "czar" within BuWeps to oversee the correction of the missile problems. Reich found this out shortly after his return from the Pacific when he was called into Stroop's office on January 17, 1962. "You've got a new job," he was told. "I am going to create a czar who is going to run the God-damned three T business," the admiral explained, "and you're it." This is how, Reich, still holding the rank of captain, became head of what became known as the "Get Well" program.[9]

Stroop gave Reich ten days to figure out how he was going to organize the thing. When Reich was done with the planning, Stroop did not want Reich to come back to him. Instead, he told Reich to take it to the special advisory group that Stroop had established to help run BuWeps.[10]

Rich talked to a number of people in OpNav and BuWeps and began to put together a plan to assemble a group of people taken from the Tartar, Terrier, and Talos technical experts within both of these organizations. In four or five days he had worked out an organization chart. He was confident enough by the end of the week to request a meeting with the advisory committee composed of retired Vice Adm. John Price, retired Adm. Frederick Withington, and a civilian inventor well known to industry by the name of Earl Canfield. Reich made his presentation using charts to show how he was going to pull together a group of people from the Research Division, the Maintenance Group, the Comptroller's Office, and so on. When he was finished with his pitch, the chair, Admiral Price, said, "It's not good enough." Price thought that Reich had taken only a small step when he should have taken a big step; he did not think Reich's plan was going to do anything. The board suggested that he tear it up and start all over again.[11]

Stroop's advisory group wanted a program more like the Navy's Special Project Office, which was responsible for creating the Polaris missile system. As Reich knew, the Special Projects Office had started from scratch with a lot of backing in the highest places and had high priorities. But Reich was dealing with a series of missiles that had already been introduced to the fleet that were deeply embedded within the various branches, directorates, and sections within BuWeps and the Bureau of Ships (BuShips). The technocrats in the shore establishment had to continue whatever they were doing, because the Navy had forty to fifty ships with the systems already installed and an equal number in various stages of construction, all of which had to be kept moving. This convinced Reich that in order to fix the problem, he had to work within the system not outside it. What he needed to do was to set up an organization within BuWeps, not separate from it or completely outside it. Reich's solution was the G-group.[12]

Around February 1, 1962, Reich went back to the advisory group for a second go, incorporating most, if not all their suggestions. They looked it over and said that it was much better, and it could be stronger, but it would work, and they would discuss it with Admiral Stroop. Stroop must have liked what he heard, because he decided that Reich needed to make a presentation to all the flag officer assistants in BuWeps. The comments after his presentation, for the most part, were not good. None of the attendees, with exception of Rear Adm. Frederick C. Ashworth, was happy with Reich's approach. Ashworth wanted the group to know that he had every reason to disagree with Reich's proposal, but he felt a very deep conviction toward and understanding of the gravity of the problem facing BuWeps. He was voting for it because something had to be done. Then Stroop stood up to close the meeting. "Gentlemen, I thank you all for being here. I am pleased that you all have stated your views in a very candid way. I appreciate that.

But I want you to know that beginning at eight o'clock on Monday morning, Eli is in business. The G-group will be instituted as he has laid out." After the meeting, Stroop provided Reich with a very strong charter telling him, "You are going to run this show and you are going to be stepping on everybody's toes. You are going to be stepping on my toes, but that's what I want you to do."[13]

After obtaining Admiral Stroop's approval to establish the G-group, Reich began to organize it into four principal units: a Terrier unit, a Tartar unit, a Talos unit, and a logistics group. Each unit was headed by a captain and staffed with two or three commanders and three of four civilian engineers. By June 1962, he had amassed a staff of around a hundred people. Along the way, he also picked up the title of special assistant to the chief of the Bureau of Ships. This came about after Reich complained to Admiral Stroop about the difficulty he was having getting the guys in BuShips to work with his group, especially in the area of radar.[14]

In late June 1962, Reich was asked to brief Secretary of the Navy Frederick Korth on the progress he was making on the 3T project. To preserve the informality and security of the discussion, the meeting was scheduled on the Fourth of July in the office of Assistant Secretary of the Navy Kenneth E. Belieu. By then, Reich had been "frocked"† by Admiral Stroop, although he would not be formally appointed as rear admiral until February 1963. In addition to Reich, Secretary of the Navy Korth, and Assistant Secretary of the Navy Belieu, Dr. Wakelin was also present for the meeting, which commenced at 0900. As the discussion began, Reich learned that although they were not critical of his performance, they were somewhat disappointed in the organization that had been established within BuWeps to handle the missile problem. In their opinion, G-group did not have the necessary stature or clout to contend with the overall size and complexity of the problem that involved so many ships. At the time, the Navy's 3T program consisted of eighty-five ships of which forty were in commission in the fleet. The remainder were in various stages of construction.[15]

When Reich had first taken on the Get Well program, there was considerable talk about creating a special project office akin to that which was

---

† "Frocked" is the term used to describe a person selected for promotion who wears the insignia of the higher grade before the official date of promotion. Although frocking does not result in an increase in pay or disciplinary power, it does bestow the respect that comes with the higher rank.

running the Polaris program. Reich was against it at the time, based on the belief that the task before him was not suited to the "nice clean-cut arrangement of a new-development program." Korth, however, felt that Reich should reconsider taking stronger steps toward organizing the group as a special project office. After several hours of discussion—just before the meeting broke up—Reich was directed to put together a memorandum outlining what his special project should look like along with a plan of action for its implementation.[16]

The document Reich prepared became the basis for the Special Navy Task Force for Surface Missile Systems (SMS); it was established by Secretary of the Navy Korth's directive of July 20, 1962. Korth appointed Reich as head of the task force with instructions to make monthly reports and presentations directly to him as to the status, problems, programs, and remedial actions that were to be taken. To facilitate Reich's ability to work within the bureaus, Korth authorized his status as assistant chief with the Bureaus of Naval Weapons, Ships, Personnel, and Supply Accounts.[17]

Shortly after the SMS was established, Secretary Korth, in consultation with Reich and Dr. Wakelin, agreed it would be helpful to provide an advisory group composed of senior representatives from the contractors who had a major role in the 3T program. What emerged from this discussion was the creation, in September 1962, of the Contractors Steering Group. It was composed of representatives from the Bell Telephone Laboratory, Western Electric, Raytheon, Westinghouse, General Electric, Sperry, Vitro, APL, General Dynamics Pomona, Northern Ordnance, Bendix, and Bath Iron Works. In addition to representatives from the various contractors, Reich also recruited a number of technical consultants as associate members. These were some of the leading technical people in APL and BuWeps. They were contracted directly to BuWeps and reported directly to Reich.[18]

One of the first problems that Reich had to deal with were issues with the SPS-39 height-finding radar and its potential replacement, the SPS-48. Neither the fleet nor BuShips, which was responsible for search radars, was happy with the recent at-sea evaluation of the SPS-39. The evaluation had revealed a number of technical problems that affected its reliability. The SPS-48, which was still in development, was designed to replace the SPS-39, which could not meet the long-range requirements of the most advanced Talos missiles. Reich considered a missile ship's search radar to be an integral part of the surface-missile system; as David Boslaugh has noted, "without the search radar's new-target designation inputs, the pencil-beam fire control radars were virtually blind." Search radar performance directly affected missile system performance. Neither Rear Adm. Ralph K. James, chief of BuShips, nor Reich was happy with the performance of the SPS-39. Reich was also aware of the lack of progress being made on developing the SPS-48 for the fleet.[19]

Toward the end of 1962, Reich asked Admiral James to set up a one-on-one meeting with BuShip's SPS-48 project manager, a civilian engineer named Donald C. Bailey. During the meeting that was subsequently held with Bailey, Reich reviewed each of the radar's problems and what Bailey was doing to correct it. Reich wanted to get the SPS-48 teamed up with the SMS project as soon as possible. What resources, he asked, would Bailey need from the SMS project office to accelerate the SPS-48 delivery schedule by eleven months. Reich wanted to install the SPS-48 and the new weapons director system (WDS) Mark 11 in the *Wainwright* (DLG 26), which was the third ship due to be commissioned, rather than in the fourth ship, the *Jouett* (DLG 29). Together, he and Bailey hammered out an accelerated delivery schedule for the SPS-48. At the end of the meeting, according the account later given by Bailey, he "confessed to Reich that he had come over prepared to do battle with yet another senior person who was out to kill his project, but was happily surprised [when] Reich replied that his only desire was to make the radar work, and to find out what he [Reich] needed to do [to] make sure that it did."[20]

A few days later, Reich had a similar one-on-one meeting with Lt. Cdr. Joseph L. Randolph, an engineering duty officer (EDO‡) who was BuShip's representative at the Univac Plant in St. Paul, Minnesota, where the WDS Mark 11 was being developed. Reich asked the same questions about the WDS Mark 11 timeline and what it would take to speed it up. "Randolph also agreed that with additional funds and the support of the SMS project he could meet the same schedule. Armed with the new schedules, Reich convinced OpNav that the SPS-48 and WDS Mark 11 could be installed first in *Wainwright*, and her ship's characteristics were thus amended."[21]

From his experience during the overhaul of *Canberra*, Reich realized that the technical groups within BuWeps were uncertain of the actions needed to fix

‡ An engineering duty officer is a restricted line officer involved in the design, maintenance, and repair of the Navy's ships and submarines and their weapons.

the flaws in the 3T missiles. Nor was fixing the flaws considered a high priority with anybody in either BuWeps or BuShips. Design defects, in Reich's opinion, were not being readily identified because of the failure to appreciate and examine the complex interplay of each system's elements. To rectify this situation, Reich formed the ad hoc SMS Technical Planning Group (later known as the Technical Planning Group I) to review all aspects of the Terrier, Tartar, and Talos guided missile systems. The group, which included members of APL and key industrial organizations selected on the basis of their background and expertise, was headed by Capt. Robert K. Irvine, technical director of the Special Navy Task Force. Irvine was detailed to the SMS project in February 1963 and was immediately dispatched to APL by Reich, with instructions to document each of the 3T systems, determine their design defects, and initiate efforts to resolve them. Under Irvine's direction, the Technical Planning Group laid out a plan for an orderly improvement and updating that envisioned an evolutionary replacement of the major missile components using versions that were compatible with the existing systems: radars, computers, guidance systems, and so on. When issued in November 1963, the "Technical Plan for the Surface Missile Systems" guided system developments through the remainder of the 1960s.[22]

Earlier, at the beginning of July 1963, Reich was walking down the hall toward his office when he ran into Cdr. Wayne Meyer. When Reich established the SMS, it was organized around three offices: one for each of the 3Ts. Each of these offices had a fire control desk, a missile desk, and a command-and-control desk to handle the specific part of the missile system assigned to each office. Meyer, whom Reich had gotten to know when Reich had gone on board the *Galveston* to assess the progress being made with respect to the Get Well program, had just reported for duty with the SMS project and was expecting to be assigned to the Talos desk based on his extensive knowledge and experience with that system.

"What are you doing here?" Reich asked.

"Admiral, I've just reported in," said Meyer. "Marilyn [the BuWeps officer detailer] told me I'd be relieving Commander Harmon Penny."

"Wait a minute," said Reich, "that doesn't sound right."

"Well, I know what I was told."

Reich told him to wait a minute while he went to the head. He then took Meyer to his office, where he asked his secretary to check on it. She looked it up and said, "Admiral, you're right again. We're sending him down to Terrier." So, Meyer reported to Cdr. Robert P. "Zeke" Foreman, head of the Terrier Fire Control desk, whom Meyer was to relieve as Terrier fire control system manager when Foreman, who had already been selected to captain, fleeted up.[23]

PART II

WAYNE MEYER

CHAPTER 12

THE V-12 PROGRAM

WAYNE MEYER ENTERED THE NAVY in 1943 under the U.S Navy's V-12 college training program. Born on April 21, 1926, he was a farm boy who grew up near Brunswick, Missouri, during the Great Depression. His father, Eugene, known as "Boob," was a hardscrabble farmer. His mother, Nettie Gunn Meyer, was a homemaker. Though the family struggled to make a living (his father lost a productive 800-acre farm in 1935, which forced the family to relocate to a much poorer 120-acre hillside plot), his parents made sure that he received a decent education. He completed grades 5 through 8 at St. Boniface, a small two-room Catholic school located not far from his home. The two teachers who ran the school were devoted to their students, according to Meyer, who later maintained, "They were determined that you were going the best you knew how to do, and if they didn't think you were doing the best you could do, little sessions occurred." After completing the 8th grade curriculum, Meyer enrolled in Brunswick High School, which he entered in the fall of 1936. It was there that he was first introduced to the Navy's V-12 Program.[1]

The V-12 Program was the Navy College Training Program designed to give officer candidates a basic education on an accelerated basis. It was established in the beginning of 1943 to meet the need to feed large numbers of college graduates into the Reserve Midshipmen's Schools that would provide the special instruction needed to prepare college graduates for service in the Navy. Getting into the V-12 Program involved a competitive exam that had to be applied for. Meyer received his application in January of his senior year. To be eligible for the exam, he was required to pass a preliminary physical exam in Kansas City. Getting to Kansas City, which was 100 miles from Brunswick, was not easy given that rationing was at its peak and few people had the gasoline needed for such a trip. To get there, Meyer had to rely on Father John J. Goetsch, the parish priest who also ran St. Boniface's school. Meyer was Goetsch's number-one assistant and had been keeping the church's books and records. As a member of the clergy, Goetsch was entitled to a "C" gasoline ration card that enabled him to buy eight gallons of gas per week. This was twice the allowance permitted the average person. At the time, no one in Meyer's class aspired to attend a university, and Meyer claims that he "didn't even know what the hell one [college degree] was." Nevertheless, the V-12 Program would have been very attractive to Meyer for a variety of reasons. After graduating from high school, Meyer was otherwise sure to be drafted. Entering the V-12 Program provided a means of avoiding the daft, since he would automatically be placed on active duty as an apprentice seaman. He would not be subject to Selective Service so long as he retained his enlisted status and kept up with his studies. In addition, the government would pay for all his college expenses, and he would even be drawing the pay of an apprentice seaman.[2]

Meyer took the three-day V-12 entrance examination in February 1943, passed, and was recalled in March or April for another physical exam in Kansas City. Because Meyer was only seventeen years old, his parents had to cosign his enlistment papers. He had to travel to Kansas City again to be sworn in, which occurred on May 12, 1943, ten days before his high school graduation. Meyer must have done well academically, for he was chosen class valedictorian. In June, Meyer received orders to report to the naval air station at Olathe, Kansas, which was the official naval institution nearest to his hometown. His orders specified that he was to bring only a small duffel bag and the clothes he was wearing, as he would immediately be issued a naval uniform upon arrival. This was not to be the case. When he showed up at the air station, the people in charge "didn't have the foggiest dammed idea what to do" with Meyer and the other V-12 men who had also been told to report to the air station. It took a day or two before Meyer and the other V-12 men were sent to the University of Kansas in Lawrence. The Navy had already established an Electrician's Mate School and a Machinist's Mate School there and were in the process of forming an administrative unit to manage the various Navy programs underway at the university.[3]

When Meyer and his fellow classmates arrived at the University of Kansas, there were no uniforms. Meyer, like the other men in the class, had to make do with what little clothing they had brought with them. "Somehow," he recalled, "pieces of uniform came in the next few days, and we just somehow got through July. We fumbled and wandered through it." To house the V-12 students, the Navy requisitioned nine fraternity houses. These were converted into barracks along with the student union, which was turned into a Navy mess hall. To feed

TABLE 12.1

V-12 CURRICULUM:* FIRST COLLEGE YEAR

	PERIODS PER WEEK**			
	1ST TERM		2ND TERM	
Mathematical Analysis I or III. II or IV (M1 or 3, 2 or 4)	5***	(5)	5***	(5)
English T-IT (E1-2)	3	(3)	3	(3)
Historical Background or Present World War. I-II (H1-2)	2	(2)	2	(2)
Physics I, II (PHI, 2)	4	(6)	4	(6)
Engineering Drawing and Descriptive Geometry (D1, 2)	2	(6)	2	(6)
Naval Organization I, II (N1, 2)	1	(1)	1	(1)
	17	(23)	17	(23)
Physical Training	2	(6)	2	(6)
	19	(29)	19	(29)

* For all students except Pre-Medical and Pre-Dental Corps candidates.

** NOTE: Figures in parentheses indicate contact hours per week in class and laboratory. Figures outside of parenthesis indicate the number of meetings per week in class and laboratory.

*** Mathematical Analysis I and II—combination course in mathematical analysis for students entering with 2 or less units of mathematics. Mathematical Analysis III and IV—algebra, trigonometry, and analytical geometry; or analytical geometry and calculus for students entering with 2½ or more units of mathematics.

Source: U.S. Navy, Bureau of Naval Personnel, Training Division, *The Navy College Training Program V-12: Curricula Schedules/Course Descriptions* (1943), 4.

the students, the Navy opened a cooks' and bakers' school. It didn't take long to get the kitchen into operation, but it took a year, according to Meyer, before anybody knew how to cook. "I mean, it was really awful." By September, Meyer had a full seabag, having been fitted for his uniform one piece at a time. He was required to carry at least seventeen credit hours of academics and nine and a half hours of physical training each week.[4]

At the end of February 1944, after Meyer had completed two terms in the V-12 Program at the University of Kansas, he was required to select a specialty that would lead either to a line position or to one of the specialized corps within the Navy. Each specialty had its own curriculum that the student would follow for the six remaining semesters in the program. The list of choices included programs for Civil Engineer Corps candidates, Construction Corps candidates, Deck candidates, Engineer candidates, Engineer Specialist candidates, Supply Corps candidates, or Pre-Chaplain Corps candidates, and Pre-Medical Corps/Pre-Dental Corps candidates. The engineering specialist curriculum was subdivided into Mechanical, Steam Engines; Mechanical, Internal Combustion Engines; Electrical, Power; and Electrical, Communications and Pre-radar. Reich chose the electrical engineering course called Communications and Pre-radar even though he did not have the "foggiest idea" what an engineer did. Meyer says it "was pre-radar that made him do it," without explaining the reason behind his interest in radar. While Meyer was in the process of selecting his specialty, the Navy decided that the various universities in the V-12 Program should each concentrate on one of the specialized curricula for the second year of the program. When Reich found out about this change, he was probably expecting to be transferred to another institution. He did not like the hot summers and cold winters in Kansas City and was most likely hoping for a better climate. He was "stunned and shocked" when Kansas was selected to be the center for electrical engineering, communications and pre-radar.[5]

The new curriculum got underway at the start of the new semester, on March 2, 1944. By then students who had selected the electrical engineering, communications and pre-radar course had transferred in, and those who had not, transferred out. Meyer ended up with two new roommates: one from Hoboken, New Jersey, and one from Snoqualmie, Washington. He spent an entire year trying to understand the guy from Hoboken! Meyer carried between nineteen and twenty credits that year, taking classes in calculus, chemistry, and engineering materials, analytical mechanics, economics, naval history, and elements of strategy, and one elective. He was also required to take two hours of physical training each week. He was in class Monday through Friday. Saturdays he would drill, drill, and drill (he never carried a real rifle, just a wooden one) until 1400, when those not on academic restriction would be released for liberty till Sunday night.[6]

As an apprentice seaman, Meyer received $50 (equal to $855 today) a month in pay. Out of this amount, $6.25 was automatically taken to purchase a saving bond, and another $6.60 was deducted for the Navy's life insurance program that he had signed up for. This left him just $37.00 a month for incidentals and off campus recreation. "You had to live by your wits," he recalled.

When liberty arrived at 1400 on Saturday, Meyer would go out and stand by the roadside in his uniform to hitch a ride to Kansas City. Somebody would always stop and pick him up. It might take a bit longer than the train, but it did not cost anything. Getting back to campus for the 2300 Sunday night bed-check was another matter. To ensure his timely return, he always purchased a return ticket for the 2000 train beforehand. Meyer claimed that he "went to Kansas City a number of times with nothing more than a few coins" in his pocket, "and came back with all of them." He could get away with this approach to liberty because of the existence of a honky-tonk joint on Highway 40, outside Kansas City, called Mary's. Servicemen did not have to pay a cover charge, and the place was filled with girls who were making good money working at North American Aviation's B-25 bomber plant. You could drink and eat all you wanted, explained Meyer, dance, and have whichever girl you could find who wanted you to go home with her. These girls were away from home for the first time in their lives, and they were living wherever they could. Most were sharing an apartment with three or more others. "By Sunday morning the apartment got pretty damned crowded." For food, Meyer would go to the Forum Cafeteria and just walk up to the counter and stand there. According to Meyer,

> You never [had to] pay. Somebody in line would come up and say, "Sailor, are you going to have breakfast here?"
>
> "Yes, ma'am."
>
> "Come right in, be my guest." So, you'd go to Kansas City and come back with the same coins in your pocket.[7]

Meyer and the rest of his class, at least those who survived, attended five days of class and half a day of drill for thirty-two months straight, except for a few short holiday periods. They would complete one term on a Friday afternoon and would draw books for the next term on Monday morning. It was very tiring. Not surprisingly, of the four hundred students or so who began the electrical engineering, communications and pre-radar course, only thirty-four made it to graduation. The rest were eliminated for one reason or another.[8]

Meyer's class was given a day off on V-E Day and a day and a half off on V-J Day, but they continued with their classes even though the war was over. The academic program continued until the last semester was completed. When the university announced the end of the program on February 1, 1946, nobody knew what to do until the Bureau of Personnel sent out an order to commission the thirty-four students who had successfully completed the electrical engineering, communications and pre-radar course upon graduation. Meyer and the other thirty-three survivors in his class would need regulation uniforms for the commissioning ceremony that was to be conducted at the same time the university would be awarding their degrees. Nobody knew what to do about it until a lieutenant sent to establish a Navy Reserve Officers Training Course at the university consulted the Bureau of Personnel manual. According to the manual, the newly minted (or about to be minted) ensigns were entitled to a uniform allowance. Meyer recalled what followed: "The 34 of us were loaded on a bus and transported to Kansas City, Missouri, 70 miles away. We were taken to the big Wolf Brothers men's clothing store, and there we were outfitted in a seabag of as much as could be available. And the Wolf Brothers did a fantastic job at outfitting us. I can't . . . remember the prices, but I think our clothing allowance was about $300.00 or something like that."[9]

On February 8, 1946, Meyer received his Bachelor of Science degree in Electrical Engineering and was commissioned as an ensign in the U.S. Naval Reserve. Since he was the first person in his family to receive a college degree, the graduation ceremony was a big deal for him and his family. To attend the event, his father drove the family all the way from Brunswick to Lawrence. Along the way, they stopped at the state line liquor store and bought a case of Three Feathers whiskey, which was a popular and cheap brand of liquor. His father knew that while you were not allowed to buy alcoholic beverages in Kansas, you could drink liquor, provided you did not create a disturbance. When they got to Lawrence, his father offered to take the family out to dinner to celebrate Meyer's graduation and commissioning. Somebody recommended an out-of-the-way joint called the Tepee, one of the few watering holes frequented by the sailors in Lawrence. By then, sixty or seventy people had joined the celebratory group. Those who could not ride in the Meyer's family car were driven to the Tepee by Meyer in a bus that had been secured by the same lieutenant who had arranged for their uniforms. As Meyer made sure to point out in later years that although the newly frocked ensigns were all underage, they managed to consume the entire case of the Three Feathers whiskey.[10]

When Meyer was commissioned, he had been in the Navy for close to three years and had never had a single day of Navy training aside from the course in Naval History and Elementary Strategy.

CHAPTER 13

FROM MIT TO DUTY AT SEA

WHILE ENS. WAYNE MEYER was sitting around waiting to receive orders for his first duty assignment after receiving his commission, an AlNav* bulletin was circulated among his classmates soliciting applications to pursue graduate training in electrical engineering at the Massachusetts Institute of Technology. None of them was sure of what to do, so they asked the ROTC lieutenant who had helped them before for advice. "Look," he said, "the Navy doesn't do stupid things. The Navy knows you've been going to school the last 32 months, and they're not about to send you back to school right now. This program will come around again, take my word for it." He recommended that they apply for the course in order to show their interest so that when it came around again, they would be "at the head of the line." It sounded pretty good to Meyer, so he filled out the application and sent it in without expecting anything to come of it.[1]

After another week had passed, Meyer finally received orders to board a transport at San Francisco for further transfer to the Pacific. He was sent by air (it was his first flight in an airplane) to San Francisco and billeted at an officer's barracks on Buchanan Street until his ship could be located. He had to report in each day, but they never did find his ship. After several days of this frustrating routine, Meyer unexpectedly received orders one morning to report to Naval Air Station Oakland for further transport to the Pacific. So, he went out to the air station and, as he recounted with some annoyance, "spent the whole damned afternoon sitting around. Gear was all loaded, everything was all buttoned up, and so 25 or 30 of us were waiting to be loaded onto this plane." While he was waiting to board the aircraft, a WAVE came running up asking for Ensign Meyer. When Meyer responded, the WAVE told him that he had a change of orders. He was to report to a place called MIT and he was supposed to be there by February 15 (it was now March 15, 1946). She also told

* An AlNav is a message sent out throughout the Navy.

Meyer that he had a "Pri 3" priority to fly with, which was high enough to be treated as "a big wheel." So, Meyer went about retrieving his gear (which "pissed off the ground crew that had to go find it") and boarded a flight to Kansas City. From there he flew to Boston. The plane landed at 1600 on March 17; he then took a cab to the MIT campus in Cambridge, Massachusetts.[2]

When Ensign Meyer started classes at MIT in March 1946, the school was overcrowded with people trying to catch up on four years of lost education. Most of the students were far better prepared than Meyer. The classrooms were so packed that there was standing room only, according to Meyer. "You just stood up, you couldn't even take notes." And there were no books. For Meyer, "it [was] just horrible."[3]

The coursework that Meyer was required to take included three semesters of X-rays and two semesters of atomic physics, both of which were far above his level of education. For example, Meyer had one class taught by Dr. Walter Müller, a German physicist noted for perfecting the Geiger counter. As Meyer explained, Professor Müller's big classroom was surrounded on four sides by large blackboards. Müller would walk into the room at the start of class, pick up a piece of chalk, reach two-thirds of the way up, which was about as high as he could reach, and start writing. He would go around the room covering each blackboard with writing, explaining as he went along. But Müller, who was born and educated in Germany, spoke horrendous English; Meyer claims that it was "incomprehensible." One wonders how Meyer managed to pass the course. But as Meyer noted later, "The idea of school and learning had been pounded into [him] by [his] mother; by Father Goetsch, the pastor of St. Boniface [and] Sister Mary Joanne." Meyer was no dummy, either. Nevertheless, things looked pretty grim for him and for the other ensigns in the electrical engineering master's program in the summer of 1946. But then, the Navy had run out of money and cut the program to three terms. "It saved our asses," declared Meyer. As a consolation prize, MIT decided to give those who had passed the course a Bachelor of Science degree in lieu of the masters that they had been sent to earn.[4]

Meyer left Cambridge, Massachusetts, on February 22, 1947, in the middle of a snowstorm, under orders to report to the radar picket destroyer *Goodrich* (DD 831). The *Goodrich* was one of twelve *Gearing*-class destroyers ordered for conversion into radar pickets in May 1945 subsequent to the Navy's experience in combating the kamikaze threat during the Leyte Campaign. A tripod mast to support the 1,700-pound, 6-foot-diameter SP radar antenna was installed forward of the No. 2 stack in place of the forward torpedo mount, which was removed. The SP was a general air and surface search set operating in the S-band that could be used for fighter direction. Its principal function was to provide range, azimuth, and altitude information on aircraft.[5]

To get to the *Goodrich*, Meyer took a bus that followed a snowplow to Newport, Rhode Island. From the bus station, he had to walk to the fleet landing, which was a good distance away, burdened with a big Val-Pak[†] bag that contained all his gear. There was no other way to get there as there was the holiday celebrating George Washington's birthday and nothing was running. When he got to the landing, there was a motor whaleboat waiting for the next mail run. They made Meyer put on a life jacket and headed out to the buoy where the *Goodrich* was moored. Meyer, in his usual graphic detail, described the trip out: "I didn't think we'd ever get there. The buoy was a long way out, and it was just storming like hell." When they reached the ship, Meyer had to climb the icy accommodation ladder.[6]

Once on deck, Meyer asked to see the officer of the deck (OOD). The captain and the executive officer were ashore. In those days, according to Meyer, the captain and executive officer did not worry very much about their ship in port and trusted their OODs. So, Meyer reported to the OOD, Ens. Edward C. "Dumbo" Atkinson. Atkinson took Meyer to the bridge (most likely after seeing that Meyer's gear was sent to his stateroom), where he spent the night learning how to steam on the buoy in order to maintain position in a storm without dislodging the buoy. Meyer was taught to steam three to six knots, depending on the surge. "The trick," Meyer explained, "was not to pull the damned buoy out of the water. . . . The idea was to try and take the pressure off of the chain, or [to] relieve the pressure somewhat."[7]

The next day, Meyer went to lunch in the wardroom—he had never been in one before—and was placed at the end of the table because everybody assumed he was "George," the nickname given to the most junior officer serving on a ship. The captain, Cdr. Leonard J. Baird, and the executive officer, Lt. Cdr. Phillip H. Teeter, were back on board and were also seated for lunch. They got to talking, and after a while Captain Baird asked Meyer a question. "Which light is the green light and which light is the red light?"

---

† Val-Pak is the brand name for a folding garment bag.

Although Meyer had been in the Navy for four years, he had never been to boot camp or gone to midshipman school, nor had he ever been on a ship; he knew nothing about seamanship. He was as green as green could be, which was quite apparent to the captain from their brief discussion. Meyer, seated at the far end of the table hesitated for a moment or two before answering.

"Well, Captain, would you repeat the question? I can't hear down here."

"Well," he said, "we have two running lights. One running light is red and one running light is green. Which is which?"

It got deadly silent, Meyer recalled. Even the stewards made sure they got far away.

"Captain, I don't know," he confessed.

Dead silence. Not a word was spoken the rest of lunch. After lunch, the executive officer called Meyer into his cabin and provided a short lesson, telling Meyer to think of port wine, which is red, and the port light is red. And that's the way Meyer remembered it.[8]

That afternoon Commander Baird and the Lieutenant Commander Teeter sat down to discuss what they were going to do with their very green ensign. The answer was to send him off to emergency shiphandling school at the fleet training center in Newport. And that was the beginning of an eighteen-month training period that provided Meyer with everything he needed to know about shiphandling and seakeeping that allowed him to qualify as OOD.[9]

When Meyer received orders to report to the *Goodrich*, his orders stated that he would be the electronic repair officer for the division, but there was no such billet in the fleet. Instead, he was to serve as the ship's radio repair officer. Once onboard, he discovered that every passageway was filled with metal boxes packed with spare parts that had been rushed on to the ship over the weekends. It turned out that all these parts belonged to Meyer. None of them was in the custody of the supply officer. According to Meyer, "There were several thousand [?] boxes of them around the ship, because the yards had just stuck them wherever they could find a place to stick them, and the inventory was terrible." Each of the boxes had a label on its exterior indicating what its contents were, but neither Meyer, nor anyone else on the ship knew where anything was. When some piece of electronics or the radar failed, Meyer had to look through all the boxes, and if he searched long enough, he might just run across what he was looking for.[10]

In addition to his duties as the radio repair officer, Meyer was also assigned as CIC officer. The division he was placed in charge of included the radarmen, the sonarmen, the radio technicians, the postal clerks, and a couple of others. Meyer's guardian angel turned out to be a little round chief petty officer named McAllister. McAllister, with a radio technician's rating, was also the chief of the division and thus served as Meyer's assistant. He was an expert, having taught at the radio school in Washington, D.C., before persuading the powers at large to let him go to sea. The former instructor started out by teaching Meyer Morse code. He once told Meyer that one "could never, never work in this kind of business without mastering the Morse code." Meyer never did fully master it, but he learned enough to make McAllister proud.[11]

Meyer was also responsible for making sure the SP-1Ms antenna located atop the tripod mast just forward of the No. 2 stack was operational. The 1,700-pound parabolic dish had to be stabilized with respect to the horizontal and the vertical pitch and roll of the ship in order for the radar to provide accurate azimuth and height readings. In those days, this was accomplished by the Mark 16 stable element that had originally been developed for the Mark 37 director.‡ The Mark 16 used a gyroscope to maintain an absolute horizontal and vertical reference, effectively creating the horizontal. Reich claimed that the Mark 16 "had a tumbler capacity of around 20 degrees" and would tumble all the time. To avoid damaging the stable element, *Goodrich* would go to sea with the antenna chained down until it was needed for an exercise.[12]

One day Lieutenant Commander Teeter told Meyer that he [Meyer] was going to stand watches. Meyer received a couple of hours of instruction, listened, and read a lot. One of the things that bothered Meyer as he was learning to stand watches on the bridge was the mindset of the watch standers who were Naval Academy graduates. Before he left the ship, Lt. (jg) Elbert S. Rawls Jr. told Meyer that "those officers who came out of the Naval Academy had developed and trained [to believe] that they were to be better on the bridge than the enlisted crew." As Meyer watched on the bridge, these "OODs would spend their time reading the semaphore, reading the flag hoist, watching for contact, doing the lookout duties, and see if they could beat the lookout at spotting a ship, or see if they could beat the signalman." Meyer did not like this one bit. When he started standing watches that summer, he called all the enlisted bridge watch standers together and said the following: "As long as I'm officer of the deck, I'm important to you to do my duties, and you're important to me. I'm

‡ See Mindell, *Between Human and Machine*, 63.

not going to do your duties! I'm *not* going to read semaphore, I'm *not* going to read flashing lights, and I'm *not* going to spend my time scanning the horizon, because there are other things I'm supposed to do as the officer of the deck." Meyer expected those under his command to do their duties without his interference, and he told them so. This approach to leadership would serve Meyer well in the years to come.[13]

Meyer's time as an OOD was limited. In September 1947, the *Goodrich* was sent to the Davisville Naval Station on the Rhode Island side of Narragansett Bay. The ship was moored at the ammunition-handling pier, which was no longer in use. To Meyer "it was like the end of the World. . . . The only way you could get out of there was by boat or bus." Because of transfers and those on leave, the crew, which normally consisted of 300 or so enlisted sailors and about 20 officers, was down to 100 enlisted and 6 or 7 officers. That was the crew status when orders were received in the later part of December 1947 to prepare to sail to Gibraltar in the Mediterranean. Recall orders urgently went out all over New England for officers and sailors to report back immediately.[14]

Before departing for the Mediterranean, *Goodrich* was sent to the New York Navy Yard where she was overhauled. On the way to the Mediterranean, she rendezvoused with the aircraft carrier *Midway* (CV 41). By then, Meyer was the senior watch stander and combat information center (CIC) officer. *Goodrich* had three contiguous CICs: one was the air CIC, one was the surface CIC, and one was the undersea CIC. They were right next to each other, and Meyer would flip back and forth between them. The CIC was right underneath the bridge and connected to it by voice tubes. It also had sound-powered phones for communication with that bridge, but for some reason the captain did not like them; instead, he would start yelling down through the voice tube wanting to know what was going on. After a while, the radarmen, fed up with all this screaming, would stuff shirts into the tubes. "The old man," as the captain was frequently referred to, Meyer recalled, "would then send down the quartermaster who would pull them out so the old man could talk to us."[15]

In addition to alert duty—in case it was necessary to evacuate civilians during the Trieste Crisis—*Goodrich* spent most of her time in the Mediterranean as *Midway*'s plane guard or taking over the task force's fighter direction duties. When not on plane duty, and if the weather was good, the ship's fuel bunkers would be topped off by *Midway*. It was Meyer's first experience in shiphandling while refueling underway.[16]

While Meyer was assigned to the *Goodrich*, the Bureau of Personnel began soliciting applications for the regular Navy.§ After seeking advice from the executive officer, Meyer wrote out a letter of application. He took it to the executive officer, who drafted an endorsement that said, "This officer shows by his record that he would be an asset to the regular Navy, and I, the captain, so recommend him." Meyer turned it in, and in a couple of months received papers to fill out. That was all he had to do to get into the regular Navy.[17]

The cruise to the Mediterranean ended on May 22, 1948. A few months later, Meyer received a message with orders to report to the *Springfield* (CL 66), a *Cleveland*-class light cruiser. In July he disembarked from the *Goodrich* for the last time and headed for Long Beach, California, where his new ship was based.[18]

§ The size of the regular Navy and the number of officers of each rank is established by Congress. Reservists (like Meyer) were expected to fill out the ranks of the Navy in times of crisis and then return to civilian jobs. Transferring to the regular Navy was a means of establishing a permanent career in the Navy.

CHAPTER 14

FROM ELECTRONICS OFFICER TO NUCLEAR WEAPONS INSTRUCTOR

WHEN ENSIGN MEYER REPORTED to the *Cleveland*-class light cruiser *Springfield* (CL 66) in July 1948, the ship was getting ready to deploy to the Western Pacific. The Bureau of Personnel determined that the ship needed an officer who knew something about electronic repair, and that is how Meyer was selected. *Springfield*'s twelve 6-inch guns were located in four turrets—two forward and two aft. Meyer claims that when he reported to the ship he was initially assigned as third division officer, which meant that he had turret three along with the catapults and the boats. This does not make sense. First, it seems likely that the division handling turret four, the aftmost turret, would have had charge of the catapults, not turret three. Second, it is hard to believe that an ensign with no ordnance experience would be assigned the responsibility for a turret, which was one-quarter of the ship's main ordnance. When chief engineer Cdr. John S. Slaughter discovered Meyer's background, however, he was reassigned as electronics officer—a billet yet to be defined by the Bureau of Personnel.[1]

The type commander of Cruiser Force, Pacific Fleet, decided that electronics repair belonged in the engineering department, which is where Meyer was assigned for duty. Besides Meyer and Lt. Richard L. Ploss, the damage control officer, the engineering department was staffed with a number of talented chief warrant officers—technical specialists who performed duties requiring strong technical competence in specific occupational fields. Chief warrant officers had authority and responsibility that was greater than that of all other noncommissioned officers, candidates, and midshipmen, but they were subordinate to the lowest officer. There were a dozen chief warrant officers, and they had their own mess. Meyer befriended these warrant officers and was frequently invited into their mess. The warrant officers, Meyer recalled, "taught me how to do things and how to get things done, and whom to respect and whom not to respect." Presumably, this was the stuff he had not learned during his destroyer duty.[2]

When Meyer arrived on board, the ship's SC-type long-range air-search radar was not working properly and nobody on the ship knew how to fix it. The SC was a search radar deployed during World War II to locate planes and surface vessels. The improved version deployed on the *Springfield* had a 15' × 15' 6" bedspring-like antenna that had two rows of six dipoles mounted in front of a common grating. Meyer claims he did not know anything about the radar at that time, but he knew a lot about the theory of electronics and decided that he would learn how to fix it. He was helped by Chief Radio Electrician Norman G. Woods, who had come on board the same day Meyer joined the ship. The two men studied the manuals devoted to the SC every night after supper for several months trying to figure out why it would suddenly stop working after a time. During the day, they would turn it on, tune it up, and begin tracking the ship's float plane that served as the test target. After an hour, it would die, and they would lose track of the plane.[3]

One night, at around 2200, they finally figured out what was wrong. A new capacitor had been introduced into the design of the receiver. It was a button-type capacitor that was fastened to the chassis using a washer, a screw, and a wingnut. Wires attached to the capacitor connected it to the radar's circuit. Working together, Meyer and Woods determined that the capacitor was defective: it would last for a while and then go to ground, causing the radar to fail. They reported their findings to Commander Slaughter and convinced him and the ship's supply officer "to order every condenser* they could order; every one they could get out of the supply system before we sailed." They got boxes full of them. According to Meyer, the *Springfield* was the only ship in the Western Pacific Fleet (WestPac) "who could find a goddamn target." It took a while for these "assholes to come to believe us," but those in charge of the receiver eventually redesigned it to eliminate the short. By the time they got around to this, Meyer and Woods had left the ship.[4]

Except for Woods (who knew nothing about radar), and two second-class radio ratings, Meyer's department was made up of strikers† and third-class

* This type of electronic component was later renamed the "capacitor."

† A striker is a nonrated enlisted man or woman officially designated as being in training for a specific petty officer rating.

ratings. Half of the strikers had just gotten out of the deck force, had not been to school, and did not know anything about electronics, radar, or radio. So Meyer, assisted by Woods, put up a couple of blackboards. Any time one of his personnel had a technical question they would go to the blackboard and hash it out. Meyer's technicians were not required to stand watches, but he changed that by establishing a watch schedule for the department.[5]

When Meyer had come on board there had been a big argument between the executive officer and Commander Slaughter about standing watches. The executive officer wanted Meyer to stand bridge watches. Slaughter insisted that Meyer stand CIC watches. Slaughter won the argument and Meyer was assigned to the CIC, which was much bigger and better equipped than the *Goodrich*'s CIC. Meyer learned a lot about CICs during his tour on the *Springfield*, which served him well in the future.[6]

The cruiser sailed for the Far East in October 1948 and arrived at the Yokosuka Naval Base in Japan on November 3. During the next six months the ship cruised with the Seventh Fleet, visiting such places as Sasebo, Yokosuka, Kure, Tsingtao, Shanghai, and Okinawa. Liberty in these ports provided Meyer with new experiences and a wealth of "sea stories" that he later shared in his oral history. He was also promoted to lieutenant junior grade in the middle of the cruise. *Springfield* returned to the West Coast on June 1, 1949. In the autumn, the ship underwent an inactivation overhaul. The mothballed ship was scheduled to join the San Francisco Group of the Pacific Reserve Fleet in January 1950. Before that happened, in December 1949, Meyer received orders to report to the destroyer tender *Sierra* (AD 18) in Norfolk, Virginia, as radio officer.[7]

The 14,00-ton destroyer tender *Sierra* was 530 feet long and had a crew of 1,050 personnel. She was designed to support eighteen destroyers and was equipped to provide repairs for up to six destroyers moored alongside at a time. In addition to repairs, the ship had training and medical facilities. When Meyer arrived on board, he discovered that there were only two other line officers besides the captain: one was the executive officer, the other was Meyer. All the others were specialists: doctors, dentists, or chaplains. In addition to being the radio officer, he became the CIC officer, the antisubmarine warfare (ASW) officer, the shore bombardment officer, and the officer responsible for registered publications and communications. While the *Sierra* was equipped with four 5-inch/38 dual-purpose enclosed gun mounts, they were provided for self-defense, not for support ashore. The fire support mission would be assigned to destroyers. Meyer's job as shore bombardment officer was to organize the fire support provided by these ships.[8]

On January 6, 1950, *Sierra* was scheduled to depart Norfolk for the Mediterranean where she would be deployed with the Sixth Fleet. As the ship was getting ready to sail, Meyer was in the radio shack when word was passed that he needed to report to the captain on the bridge, which was 30 or 40 feet from the radio shack. Meyer hurried to the bridge. When he got there the executive officer, Cdr. John A. Hack, was standing there with the skipper, Capt. Eugene C. Rook. Hack turned to Meyer.

"Look, you came out of a destroyer, didn't you?"

"Yes, sir," Meyer quickly replied.

"And a cruiser?" Hack asked.

"Yes," answered Meyer.

"Well, you're a qualified officer of the deck, aren't you?"

Meyer responded again, this time with more detail.

"Yes. I've stood a number of watches in the destroyer. My watch standing in the cruiser is limited, but, as a matter of fact, I stood as the senior watch officer in the destroyer."

That was all the two senior officers needed to know. They turned to each other, and Captain Rook said, "Get her under way." And that was that. During the nine-day passage to Gibraltar, Meyer alternated watch duties with Commander Hack until they could get somebody else qualified. Thus, Meyer became the sea detail OOD and the maneuvering OOD and had to be on the bridge in any kind of formation when Commander Hack was not on the bridge.[9]

Meyer made two cruises to the Mediterranean on the *Sierra*. By the end of the second cruise, he was fed up and was applying for any duty that would get him off the ship. Nothing angered him more than the stuff they had to do for the admiral in command of the Sixth Fleet. In those days, the tender played a big role in providing logistic support to the flag officer in command. The admiral during Meyer's cruises was Vice Adm. John J. Ballentine. Admiral Ballentine and his staff were on a cruiser, which would go into port first. The *Sierra* would be about fourth in line. Meyer's description of what followed clearly shows his dislike of what the watch officers had to put up with as they were trying to enter the harbor: "Already the goddamn blinking lights were just going like crazy: Where's the barge? Where's the cars? On and on and on. Christ, we were still trying to get up the channel. That was a rough undertaking. That was hard work. And, of course, they'd get pissed off at you, figuring you're unresponsive. Staff people can be really snotty. They can really be snotty."[10]

Meyer was married by this time and was the father of a little girl. He loved the Navy, and he didn't want to leave it, but everything seemed "just so damned haphazard." On the way back to port during his second cruise, Meyer was notified that he had been selected to go to the Guided Missile School in Fort Bliss, Texas, in August. Meyer had developed an interest in guided missiles after reading about them in a number of magazines, and the school was one of the assignments he had applied for. Meyer and his wife were tickled about his new assignment because it was not another ship; but it created problems, too, because it required his wife and child to relocate again. His wife's family was not happy that she had married a "sailor boy," and they kept telling her she was never going to have anything but a haphazard life, that she would just be constantly moving. To his wife's credit, she ignored their displeasure and made the most of the situation.[11]

During World War II, Fort Bliss had been the U.S. Army's Antiaircraft Training Center. When Meyer reported for duty in August 1951, the guided missile training program known as the Guided Missile School was run by the Army Air Defense Center, established there by the War Department in July 1946. The Army and the Navy were both developing a wide range of guided missiles by then, and the school was established to develop a pool of officers for future assignment, when the new missiles became operational. These officers needed to be sufficiently familiar with the technical details of guided missiles to be able to evaluate the new missiles and make tactical studies for the integration of guided missiles into their respective services. Fort Bliss was the ideal location to undertake this mission because of the sparse population that surrounded it and the abundance of mountain-ringed desert that was ideal for rocket and missile experimental firings. The mild weather permitted year-round field-work, firing, and testing.[12]

Meyer did not know it, but the course that he was about to take put him on the cutting edge of a technological revolution that would change the nature of the U.S. Fleet. He was on the ground floor of the changes that would take place. He had no idea of the path upon which he was about to embark.

Meyer's class consisted of ten Navy officers, ten Army officers, ten Air Force officers, and five Marine officers. Most were lieutenants, and there was a captain or two among them as well. In addition to classroom work, the students were frequently taken to the McGregor Missile Range or the White Sands Proving Ground, where they observed test firings and participated in practice missile firings. During one of these exercises Meyer was on the firing team for the Navy's Lark missile. The design of the 1,210-pound missile had been initiated by Bureau of Aeronautics in 1945 for defense against the kamikaze attacks. The missile, which was developed and built by Fairchild Aviation, was armed with a 100-pound high-explosive warhead and was detonated by a radar proximity fuze. The missile used radio-command mid-course guidance and semi-active radar developed by the Raytheon Company for terminal homing. The take-off from a zero-length launching cradle was made with two solid-propellant boosters housed in a jettisonable box-fin structure. On January 13, 1950, according to a memorandum by Rear Adm. Grover B. H. Hall addressed to the CNO, a Lark test missile with a Raytheon APN-23 homing radar made the first successful aircraft interception by a U.S. guided missile. Further development of the subsonic Lark was terminated in late 1950 because of the work being done on the Bumblebee program managed by the Applied Physics Laboratory (APL).[13]

After seven months of classroom lectures and instruction, the teachers running Meyer's course ran out of material. For the next five months, Meyer and his classmates traveled around the country visiting aerospace plants and laboratories. They visited such companies as North American Aviation in Lakewood, California, and they spent several weeks at China Lake—the Naval Ordnance Test Station near Inyokern, California. During a visit to Convair in San Diego, Meyer had a chance to meet Alex Kossiakoff, the supervisor of APL's Launching Group (BBL). APL, which was owned and administered by Johns Hopkins University, had begun developing supersonic ship-launched anti-aircraft missiles in January 1945 at the bequest of the Bureau of Ordnance. Kossiakoff's BBL, working with the Hercules Powder Company, had developed large single-grain solid fuel rocket motors that were used as boosters for the STV-3 test rocket. These STV-3 test rockets were used by APL to explore the behavior of prototype guidance and control systems at supersonic speeds. The success of the STV-3 led the Navy to suggest that APL convert the rocket into a short-range guided missile that would become the Terrier—one of the so called 3Ts that APL would develop. Convair was selected to build the missile, and Kossiakoff was there to help with its development. Both the Terrier and APL would play an important role in Meyer's future.[14]

After finishing the guided missile course at Fort Bliss, Meyer was sent to the Sandia National Laboratories in Los Alamos, New Mexico, to prepare him for duty as an instructor in nuclear weapons at the Fleet Training Center in Norfolk, Virginia. Meyer was not pleased with the assignment, if not depressed. Every junior officer that he knew who had gone into special weapons stayed there. It looked like a dead end career-wise.[15]

During his three months at Sandia, which manufactured all the U.S. nuclear weapons, Meyer was taught everything there was to know about the Mark 6 atomic bomb that was rapidly replacing the Mark 4, which was an improved version of the Fat Man/Mk3 that was similar to the bomb dropped on Nagasaki. The Mark 4 was the first atomic bomb to be produced on an assembly line and stockpiled in large numbers. When Meyer began teaching a course in atomic weapons in April 1952, the only Navy aircraft that could deliver the 7,600-to-8,500-pound Mark 6 was the AJ-1 Savage. It was the first aircraft designed specifically to carry a nuclear bomb. A much lighter weapon, the B5, which weighed only 3,000 pounds, was designed for use by tactical aircraft and would enter service a year later year.[16]

Meyer left Sandia at the conclusion of the training program on atomic weapons and reported to the Fleet Training Center in Norfolk, where he and seven other officers established a school to train and orient naval officers and civilians on atomic weapons. One of their most important tasks during the eight-to-nine-week course was to instruct the Navy's people on how to assemble and employ atomic weapons. After graduating from this course, the officers who took it were certified by the Navy and the Armed Forces Special Weapons project to handle nuclear weapons. Several of the people taught by Meyer ended up in special weapons units (SWUs) that were responsible for assembling atomic bombs. Meyer and his fellow instructors also conducted three-day orientation classes specifically designed to accommodate sixty to eighty very senior people. This orientation course would generally be divided into half-day sessions: one devoted to how atomic bombs work and how they are made; one devoted to target selection and employment; and a third devoted to the Atomic Energy Commission and the national organization, and so on. There would also be talks by the unit's commanding officer. Meyer taught the course seventy-six times and claims that he never tired of it because he changed it every time. About half the staff got to see the test firing of a nuclear explosion in Nevada. Meyer was scheduled to go but got shut out and never did see a nuclear weapon's test.[17]

Meyer turned out to be a pretty good teacher and was complimented on the courses he taught, information that undoubtedly made its way to his commanding officer, Cdr. Harry Marvinsmith. By then Meyer had begun seriously to think about his future in the Navy. He knew that getting promoted was highly competitive and that he did not have the credentials needed to move up the line. He was particularly concerned with his lack of knowledge regarding navigation, for which he had no formal schooling. Meyer discussed his concerns with his commanding officer. Commander Marvinsmith, an ex-reservist himself, had attended the General Line School. Fortuitously for Meyer, Marvinsmith's good friend, Capt. Allen P. Calvert, was chief of officer detailing in the Bureau of Personnel. Marvinsmith convinced Calvert to cut orders to send Meyer to the General Line School for his next duty assignment.

CHAPTER 15

GRADUATE SCHOOLS AND MORE SEA DUTY

MEYER'S SELECTION TO ATTEND the General Line School in September 1954 was a choice assignment for a junior lieutenant. Established in 1946, the General Line School had been set up to prepare non–Naval Academy graduates for assignment on board ships. In 1951 it moved from Annapolis, Maryland, to Monterey, California. The curriculum when the school opened was divided into five major divisions, subdivided into the individual courses as follows:

I. Operational Command
   1. Strategy and Tactics
   2. Communications
   3. Combat Information Center
   4. Antisubmarine Warfare
   5. Aviation

II. Administrative Command
   1. Naval History
   2. Naval Law and Discipline
   3. Foundations of National Power
   4. Logistics
   5. Intelligence

III. Ordnance and Fire Control
   1. Ordnance
   2. Fire Control

IV. Engineering and Damage Control
   1. Engineering
   2. Electricity and Electronics
   3. Damage Control
   4. Mathematics (which was a review course)

V. Seamanship and Navigation
    1. Seamanship
    2. Navigation
    3. Meteorology

The eleven-month course involved nearly 1,300 hours of instruction, which, in addition to actual class time, included laboratory studies and weekly periods of review. Between 10 and 30 percent of the instruction was devoted to training on actual operating gear and mock-ups.[1]

Meyer sailed through program. "I had no problems at all," he recalled. "Hell, I didn't have to study or do anything. As a matter of fact, by the third month, I was doing a lot of the teaching myself." The courses he enjoyed most were the maneuvering board and navigation. He liked leading one of the teams and learned more about maneuvering and navigation than he thought he would ever have the opportunity to learn.[2]

After completing the course of instruction at the General Line School in May 1955, Meyer was ordered to the radar picket ship *Strickland* (DER 333) for duty as the executive officer. Meyer's selection for assignment to a radar picket was a natural choice given his previous experience as an electronics officer.

Strickland (DE 333) had been a World War II destroyer escort; it was converted into a radar picket by the addition of an SPS-6B long-range air-search radar, an SPS-8 height-finder, and an SPS-4 surface/zenith radar. The latter was a high-resolution, short-range surface-search and zenith-search radar for use on board naval vessels of moderate size. It had a dual antenna assembly that permitted the operator to observe targets either on and near the surface of the water or approaching from overhead. The converted radar picket had an enlarged CIC with four radar oscilloscopes that enabled the ship to direct up to four airborne intercepts simultaneously. *Strickland* worked hand in hand with the U.S. Air Force, which maintained a network of radar stations along the Atlantic Seaboard. Operating from her homeport in Newport, Rhode Island, the ship served at various picket stations on the North Atlantic Seaboard.[3]

In addition to his duties as executive officer, Meyer was also the ship's navigator and fighter direction officer. Part of the ship's mission was to test the communication setup mandated by headquarters staff on the Atlantic Fleet. Although the Air Force was responsible for the air defense of the continental United States, the commander-in-chief of the Atlantic Fleet (CinCLant) insisted that all naval communications be routed through his headquarters. To accomplish this task, *Strickland* needed to have three speed-key certified radio operators on board to send the unremitting ship reports via continuous-wave Morse code. The communications to CinCLant headquarters were so bad during his first trip that the patrol had to be cut short. As they headed back to Newport, the ship was diverted enroute to the Philadelphia Navy Yard where a new SRC-16 high capacity, long-range high-frequency communication system consisting of eight large cabinets was installed. "They were pretty big suckers," Meyer recalled. Once the equipment was installed, the ship had no trouble getting its reports through to Norfolk.[4]

Meyer considered the *Strickland*'s patrols in the North Atlantic the toughest sailing he had ever done. He complained, "You never had anything warm to eat, you couldn't stay in your bunk, you couldn't stand up. You spent your whole time struggling." Crowding was pervasive, and the wardroom was so small that two shifts were required. The weather was so bad that lifelines were continually rigged all over the ship, and no one was allowed on deck at night without special permission from the officer of the deck.[5]

The turnaround time for the ship from port to port was something like twenty-seven days, which was possible only because the *Strickland* was diesel powered and could get underway without the need to make steam. Meyer remembered, "It would be three days out, three days back, and twenty-one days on patrol." Conservation of fuel was always a significant factor, hence the reason the Navy chose to convert its diesel-powered destroyer escorts to radar pickets for this particular mission. One downside constantly mentioned by Meyer was the smell. He would go into the officer's club in Newport and the officers sitting there would turn to one another saying, "Do you smell something funny in here?" It was not long before he stopped going to the officer's club after the day's work was done. And when he returned home, the first thing out of his wife's mouth would be, "What's that smell?"[6]

Ever since his posting to MIT was terminated by the Navy for lack of funding, Meyer was determined to get back to school so he could obtain a master's degree, which he considered "a point of pride." But whenever he requested such duty, the staff at the Bureau of Personnel would tell Meyer that he had already been to graduate school. A chance meeting with Cdr. John R. Wadleigh, the commodore of Destroyer Escort Squadron 16, to which *Strickland* was assigned, was in a circuitous way responsible for Meyer earning his master's degree.[7]

Commander Wadleigh enjoyed visiting the ship when it was in Newport. According to Meyer, he just liked to come and sit in the wardroom and talk with the ship's officers. Meyer explained, "You just wanted to gather around and listen to him because he was so calm and straightforward and spoke well." Meyer must have made a good impression on Wadleigh, for he was the person that Meyer claims was responsible for recommending him for duty on the staff of the commander-in-chief of the Atlantic Fleet, which ultimately paved the way for Meyer's return to MIT.[8]

One day, while the *Strickland* was in Newport, Capt. John N. Shaffer, the CinCLant's assistant chief of operations sent for Meyer to discuss his assignment to the staff as the assistant plans and operations officer. After interviewing Meyer, Shaffer took him to see the chief of staff, Capt. Benedict J. Semmes Jr. Meyer was not thrilled with the idea, but he told Semmes that if joining the staff would get him back to graduate school, he would take the job. Semmes said he would think about it. It was not long, however, before Meyer received orders to report to CinCLant's staff. Meyer had elicited a tentative agreement with Semmes that would allow him to go back to graduate school at the end of his staff tour. When Meyer submitted his application to Bureau of Personnel requesting graduate school as his next posting, both offices provided extremely favorable endorsements, which Meyer claims got him there.[9]

Meyer had no idea what he was getting into when he joined CinCLant's staff. Instead of scheduling ships, which was taken care of by the operations officer, he was assigned the job of handling the shore establishment. This task included oversight of the construction contract for a new Pier 1 that was being built to replace one lost to a recent hurricane. A chief petty officer (a boatswain's mate on shore duty) was assigned to assist Meyer. The two were in charge of getting the pier fixed and were given all the "shitty little" shore jobs that had to be done.

Meyer recalled one incident when his morning started at 5:30 a.m. with an urgent phone call: "Lieutenant, this is the chief. You'd better get down to the pier, and you'd better get down here, like right now."

By the time Reich got down to the pier, a reserve lieutenant (jg) who was assigned to public relations duty had already been talking to a reporter from the *Providence Journal*. The reporter had gotten wind of an oil spill and wanted more information. The jg who answered the phone told the reporter, "If you want to see the biggest oil spill you've ever seen, you'd better hurry up and get down here."[10]

The oil spill, which amounted to 11,000 gallons of oil, had been released from the *Forrest Sherman* (DD 931) when a stripping tank began overflowing sometime during the midwatch. It was not detected until the early morning hours, when the security watch making his morning rounds looked over the side and saw oil in the water. When the ship's CO, Cdr. Russell S. Crenshaw Jr., arrived at 0800, according to Meyer, he calmly ordered the lines taken in, got the ship underway, and left the "goddamn" mess behind them. Meyer got stuck cleaning up the mess, which included a pile of paperwork, along with the public relations effort that had to be put in place to deal with the press.[11]

Not all of Meyer's duties as assistant planning officer were without professional rewards, however. Maintaining the war plan was a big part of his duties. He had to keep track of where every ship was and its status in case they were mobilized. As it turned out, the knowledge he had obtained in atomic weapons was a valuable asset that enabled him to make expert contributions to the staff's war-planning efforts. "They were in awe of what I knew about these goddamn things that none of them knew anything about," explained Meyer. "So I became kind of authoritative on that subject."

When Meyer was nearing the end of his tour with CinCLant, he duly received orders to report to the Naval Postgraduate School in Monterey, California, which he entered in June 1958. By then he had been promoted to lieutenant commander and automatically became a section leader. Everyone else in his class was a lieutenant, so Meyer outranked everyone. As the section leader he had to make sure that all the reports were in on time, and when the class had to do something, or needed to be mustered, the order went through Meyer. It was more of a prestige post, according to Meyer's recollection, but it was also a lot of work.

The course work was tough. The general ordnance curriculum was the same for the first year and a half. Then the class was divided into groups of six that were, in turn, assigned to a specialized branch of study such as fire control, chemistry (that covered warheads and propulsion), aeronautics, and so on. Which specialty Meyer was placed in remains unknown, but at the end of two years, he still did not have enough credits for a graduate degree. What was worse, he was not one of those in the class selected for further schooling.[12]

So, it looked like Meyer was never going to get a master's degree. And then lady luck appeared once again to rescue Meyer from the intransigent detailers in the Bureau of Personnel. The Navy had paid tuition for six officers to attend MIT's graduate school. Meyer had already been designated for another duty, but one of the six officers chosen to fill the Navy's quota to attend MIT was detailed for sea duty. This left Meyer as the only officer who could meet MIT's qualifications, so the Bureau of Personnel had no choice other than to send him back to MIT.[13]

Meyer, at thirty-five years of age, was the oldest student in the class, which consisted of the six naval officers and about twenty civilians, many of whom were graduate students in their early twenties. The Navy guys, who knew how to get organized, would parcel out the work, sharing the information needed to finish the assignments given out by the instructors. Monterey, according to Meyer, had prepared them well for graduate work, and studying together made it a lot easier (the twenty-two and twenty-three-year-olds, Meyer observed, "had to work their asses off"). Meyer's thesis adviser, Professor Robert K. Mueller, treated them like adults. He felt that they were old enough to know what was worthwhile and believed that whatever they did would be meaningful. He gave Meyer an "A" on his thesis,* which along with the other grades he received that year, made Meyer a grade "A" student—the only time in all his schooling that ever happened.[14]

---

* Meyer's thesis, coauthored with Lt. Thomas I. Noble, was titled "An Analysis of Precepts Available to Synthesize Feedback Control Systems when Output Characteristics Are Specified" (master's thesis, MIT, Department of Aeronautics and Astronautics, 1961).

CHAPTER 16

TALOS FIRE CONTROL OFFICER

MEYER RECEIVED HIS MASTER'S DEGREE in aeronautical engineering from MIT in June 1961. His master's degree, coupled with his previous training in guided missiles and nuclear weapons,* as well as his recent promotion to lieutenant commander, made him the ideal candidate to relieve the fire control officer on board the guided missile cruiser *Galveston* (CLG 3). The ship was the first of the three Talos-armed guided missile cruisers in commission when Meyer joined the ship in July 1961. It was a plum assignment for a newly promoted lieutenant commander. He would oversee the firing of the long-range Talos, the newest surface-to-air guided missile in the Navy's growing stable of guided missiles. And it was a significant assignment, given the importance of Talos and the emphasis the Navy was placing on the development of guided missile systems. Meyer was initially ordered to report to the Fleet Anti-Aircraft Training Center at Dam Neck, Virginia, for orientation on Talos; but his orientation was canceled by the urgent need to replace *Galveston*'s fire control officer, who was leaving the ship.[1]

Galveston was a World War II *Cleveland*-class light cruiser that had been converted to handle the Talos missile system. To accommodate the guided missile battery, her aft turrets, 5-inch gun house, and seaplane catapults were removed, and the superstructure was modified to provide space for the missile magazine, a missile handling area, and barbettes for the SPG-49 guidance and the SPW-2 beam-generating radars. The Mark 7 guided missile launching system (GMLS) consisting of a single dual-rail launcher, a missile handling area above the main deck for 16 ready rounds, and a 30-round magazine were installed aft. Due to their large size (32 feet in length) Talos missiles had to be stowed horizontally

* The nuclear-armed version of Talos, designated SAM-N-6BW1, had recently entered service. Meyer's familiarity and security clearance with "special weapons" would have been an added asset in evaluating his qualifications for Talos duty.

in a compartment 36 feet long and 28 feet wide. Horizontal stowage provided a more economical use of space and was less complex mechanically than the vertical stowage used for Terrier. In addition to the GMLS, additional space had to be created for the radar consoles, the fire control computers, and a weapon control system. The fire control system was a functional copy of the ground system used during development testing at White Sands Missile Range. The Talos Mark 2 weapons director system (WDS), which was part of the Mark 77 fire control system, could track up to six targets at a time, assigning any two of them to fire control channels. The Mark 77 utilized the Mark 111 electromechanical analog computer to evaluate and designate targets, to position launchers and guidance radar, and to provide programming for missile guidance. A Talos system required 57 seconds to fire the first salvo and 46 seconds for each subsequent salvo.[2]

The fire control computers used on these early Talos ships were extensions of the fire control technology developed in the early 1950s. For the most part, they depended on electromechanical analog devices that were subject to a long setting time and inaccuracies caused by drifting. The drifts, according to APL's Elmer Robinson, "were not detectable without comprehensive tests and analysis. As a result, extensive test and adjustment procedures were developed." The

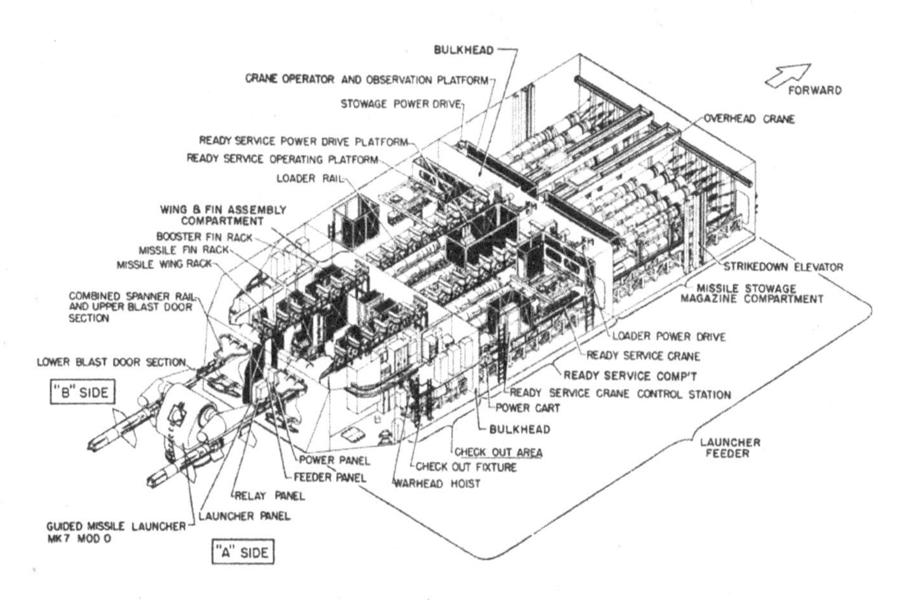

FIGURE 16.1. TALOS MK 77 MISSILE LAUNCHING SYSTEM

Source: U.S. Navy

electromechanical computers suffered from a high rate of failure that increased the maintenance burden and downtime of the fire control system.[3]

The Talos SAM-N-6b (Talos 6b) missiles placed on board the *Galveston* were 19 feet long, and each weighed 3,185 pounds without its booster, which added 11 feet in length and 4,425 pounds to the overall size of the missile. Talos was so big that it dwarfed a 16-inch battleship projectile inspiring one officer to remark, "I can imagine that the gunnery and supply officers of the USS *Missouri* (BB 63) would have very little hair left after a tour of duty with this weapon." It was twice the length and weight of a Mark 15 torpedo, the largest piece of ordnance heretofore placed on board any U.S. Navy ship.[4]

Talos 6b flew at Mach 1.8, had an effective range of 50 nautical miles, and could engage targets at altitudes up to 60,000 feet. It employed pulse-radar semiactive homing, and its 300-pound conventional warhead had a kill radius of approximately 70 feet. The Talos 6b missile was propelled into the air by a 4,425-pound solid-fuel Mark 11 booster that burned for 5.2 seconds before dropping away. By then the missile was eight miles downrange and had attained a speed of Mach 2.5. During this boost phase, the Talos 6b was actively stabilized (to compensate for the aerodynamically unstable booster-missile configuration) by command signals from a free gyroscope (uncaged just prior to launch)

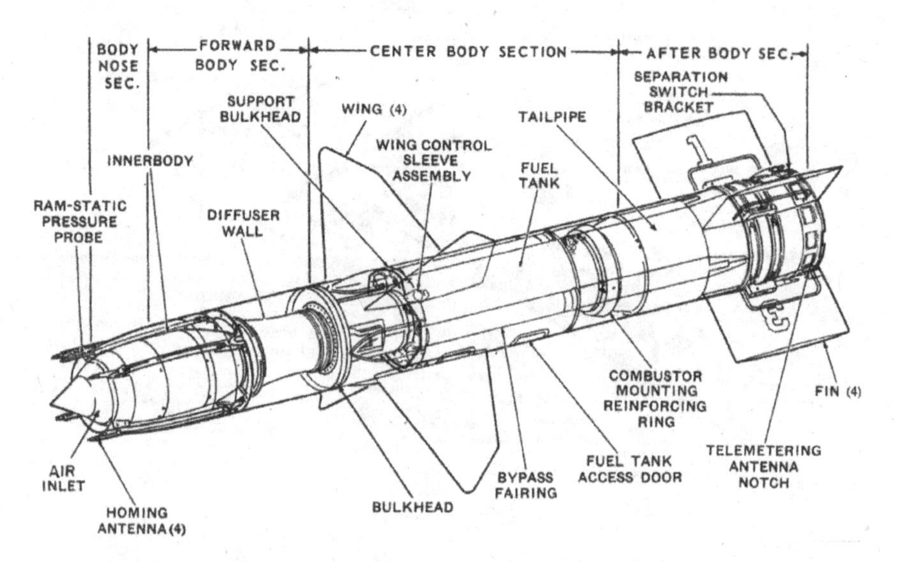

FIGURE 16.2. TALOS MISSILE
Source: U.S. Navy

that maintained the missile's orientation in space via its steering planes. Rate gyroscopes were used in each of the wing planes to provide damping.[5]

After separation from the booster, the ramjet engine was ignited. At this point, the missile was aerodynamically stable, and control was switched from the attitude stabilization mode to a midcourse beam-riding mode. An error signal, proportional to the angular off-beam error, was generated by the beam-rider receiver and fed to the appropriate steering command channels. Velocity was controlled by a Mach-sensing fuel-control system.

As the flight progressed, the direction of the guidance beam was programmed by the fire control computer to cause the missile to fly the desired midcourse trajectory. The missile's range was determined by tracking its onboard beacon that emitted radio pulses. That range was used by the fire control computers to control the beam and to compute the time at which a homing-enabled pulse code was transmitted to the missile, activating the terminal homing system. After the missile had been placed in a favorable position relative to its target, a signal from the guidance transmitter activated the terminal phase of guidance. Target acquisition was normally completed soon enough after activation to permit at least ten seconds of homing before intercept. This was enough time to eliminate missile heading errors and to compensate for target maneuvers.[6]

When Meyer joined the *Galveston* at the end of July 1961, the ship had just completed the last phase of the initial Talos evaluation trials. These were conducted in the Caribbean and included extensive testing of the missile system with multiple missile firings. After the trials, the ship briefly visited Bayonne, New Jersey, where her fire control radars (presumably the SPW-1 prototypes) were removed in preparation for additional conversion work that was scheduled to take place in the Philadelphia Naval Shipyard beginning in September.[7]

Galveston remained in the shipyard for a full year, less one month, during which time extensive changes were made to the ship and to the Talos fire control system. The changes were necessary to accommodate the 100-nautical-mile range of the extended range upgraded Talos SAM-N-6bl[†] missile. These changes involved increasing the tracking range of the fire control radar, increasing the instrumented range of the fire control computer channels,

---

† This upgraded Talos 6bl was redesignated RIM-8B when the Joint Designation System was introduced in 1963.

and adding a low-noise-level high-power continuous wave (CW) target illuminator. Modifications to the ship included the addition of missile control spaces such as those installed later on the *Oklahoma City* (CLG 5) that included a Weapons Control Center in the aft superstructure, a control room for the SPG-49 radar located deep in the ship above the Mk 111 missile fire control computer room.[8]

On July 1, 1962, while *Galveston* was still being overhauled, Meyer was promoted to commander. This was an important step up given his responsibilities as weapons officer. The eleven months that *Galveston* remained in the Philadelphia Navy Yard gave Meyer enough time to become familiar with all aspects of the Talos missile system. As the ship's weapons officer, he would have been directly responsible for overseeing that the new fire control system was properly installed and operating correctly—to the extent that this could be determined from shore. One of the biggest issues Meyer faced during his tour as weapons officer and "gun boss" was the problem of trying to communicate with the other senior officers on the *Galveston* when the ship was prepared to fight at General Quarters. If the guided missile cruiser had been designed from the ground up, it is likely that more thought would have gone into the need to provide better communication between those officers in command and control. But as Meyer noted, the engineers converting the *Galveston* "had a design job to do that old business of putting ten pounds in a five-pound bucket." Except for the Talos Weapon Control Center located in the aft superstructure, the rest of the ship was operated using the original command and control spaces designed in World War II for what was essentially a gun platform. At General Quarters, Meyer, the missile fire control/weapons officer, was located in the Weapons Control Center in the aft superstructure; his assistant, in charge of the guns, was forward up in the 06 level; the captain was also forward, on the bridge at the 05 level; and the operations officer was in the CIC, located deep in the bowels of the ship. "Communications," recalled Meyer, "was a real bear, a real bear. You had the 21MC,[‡] and you had the sound-powered telephones.[§] If you set out to go talk to each other, you'd never make it. The war would be over before you got there, because you couldn't move in the ship; you were just locked in once you went to general quarters."[9]

‡ The 21MC was the captain's command communication circuit.

§ Sound-powered telephones were an emergency shipboard communication system that relied on the energy generated by the user's voice.

Meyer found the flight time of the Talos missile the most unnerving part of his job. The four to five minutes it took the missile to reach its long-range target was the hardest thing that Meyer had to deal with in all his missile experience. He smoked cigarettes in those days and would light the smoking lamp while the missile was flying out, which left plenty of time for members of the firing crew and himself to smoke. "Then you just sat there atremble the whole time," he explained, "because our expectation was that it would make it."[10]

As the ship's weapons officer, Meyer was also responsible for the GMLS, which comprised the missile stowage magazine, a ready service magazine, a wing and fin assembly area, missile test cells, a warhead magazine, the launcher, and numerous pieces of machinery to move the missiles, boosters, and warheads throughout the missile house. A complex overhead crane, elevator, and "rammer" system was used to move Talos missile components from storage, along the assembly line, and out onto the launcher's rails. Test cells were located on the port and starboard side of the wing and fin section. They contained hydraulic and electrical power supplies for the missile under test. During testing, the missile electronic and hydraulic systems were powered up and checked to verify that everything worked correctly.[11]

Meyer claimed that he "fired something like seventy-three Talos missiles"[¶] in his life, more than anyone else. He also claimed to have been the first person to hit a surface target with a surface-to-air missile. According to Andreas Pasch, all Talos variants had a surface-to-surface engagement capability. Although the Talos 6bl on board the *Galveston* could have been used as an anti-ship weapon, its range was limited to the radar horizon because of its reliance on beam-riding for guidance. Its continuous rod warhead—designed to destroy an aircraft—would not have been very effective against ships, either. Practical use of Talos 6bl as an anti-ship missile became a reality only when a new version of the Talos, known is Unified Talos (SAM-N-61), entered service in 1962. The new missile had a CW seeker, solid-state electronics, and an interchangeable high-explosive or nuclear warhead. The new missile's CW interferometer guidance gave it the capability to engage surface targets well

¶ This was one-third of all the Talos RIM 8B-C-D missiles fired from the seven Talos guided missile cruisers in service form 1961 through 1979 (see the firing table compiled by Hays in "Talos Missile History"). The number of missiles fired quoted by Meyer during his oral history interview is suspect, given that it represents what would have been almost two full loadouts of *Galveston*'s magazine.

beyond the radar horizon with only minor changes required to the shipboard fire control system.[12]

The Talos 6bl missile fired at a surface target in a test conducted by *Galveston*, according to Meyer's account, was an experiment to see if it could be used as an anti-ship weapon. The first Talos 6bl fired for this purpose was launched against a decommissioned destroyer escort that was adrift; but it missed the target. Meyer sent word back to Washington, which gave permission for a second try. Meyer had a second missile readied and set up in the special service surface mode. It was late in the afternoon, near sunset before the *Galveston* was ready to try again. When the second missile was fired, the target was twenty-four miles downrange. Meyer was confident that he hit the target, but he would not know for sure until the ship, which raced ahead full speed, reached it an hour later. Meyer was exhilarated when they finally reached the target. "She was afire," he recalled, "just burning like hell." Although the Talos 6bl missile had not been armed with a warhead, the residual fuel—and there was a lot left after the short flight—"tore the living hell out of the ship."[13]

Cdr. Wayne Meyer left the *Galveston* in July 1963 with orders to report to the Special Navy Task Force for Surface Missile Systems (SMS) within the Bureau of Weapons in Washington, D.C. How Meyer was chosen for this posting remains unknown, but his appointment to Reich's SMS was a no-brainer. He was one of a select group of officers who had served as weapons officer on one of the few guided missile ships in the Navy. He had fired more Talos missiles—he claimed—than anyone else. Meyer also held a master's degree in aeronautical engineering from MIT. It was during his service on board the *Galveston* that "the aura of being an ordnance person got created." At that point in his career, he gave up the idea of working toward a major ship command and became a missileer. Meyer's decision to become a missileer is understandable, given his miserable experience at sea and the depth of his technical training and knowledge. He probably realized that he had a much better chance of promotion in this area than in trying to obtain an important sea command, which was very unlikely given his service record to date.[14]

PART III

ADVANCED WEAPONS SYSTEMS

CHAPTER 17

ELI REICH, DIRECTOR, ADVANCED SURFACE MISSILE SYSTEM

AS THE READER WILL RECALL, in July 1963, when Wayne Meyer bumped into Eli Reich on his way to sign in at his new posting, Reich had been knee-deep in trying to fix the problems with the 3Ts. Early on, Reich realized that a major factor contributing to the ongoing missile problems was the lack of coordination between the various engineering groups responsible for the different components and subsystems that made up each missile system. As David Boslaugh has noted, "At the time, responsibilities for developing surface missile system components such as the missile, propulsion, fire control system, weapons direction system, and the launching system were each under a different R&D manager who had his own separate development fund. BuWeps then used a system to 'interface correlation drawings' and a change control board to pull the various systems together."[1]

One of the ideas that emerged from Reich's leadership of the Get Well program was the need to establish an organization to take responsibility for overall system engineering. To this end he created the Naval Ship Missile Systems Engineering Station (NSMSES) that was established at Point Hueneme, California, on July 1, 1963, to provide in-house engineering and management support for surface missile systems. NSMSES became responsible for system integration and the coordination of instrumentation, test check out, and waterfront engineering support during the Ship Qualification Trials that were conducted on all guided missile ships immediately after a shipyard overhaul. To accomplish this task, Reich established the Ship Qualification Assistance Team, composed of an officer in charge, a project supervisor, and at least one expert technician for each major piece of equipment. The Ship Qualification Assistance Team supervised the ship's personnel during equipment tests conducted to bring the guided missile systems up to a state of operational readiness. They also trained a ship's personnel in the operation, maintenance, and command and control of the missile systems, with the objective of leaving behind a fully trained crew ready to take over when the Ship Qualification Assistance Team left.[2]

Within a year and a half, the Get Well program under Reich's leadership identified and corrected many of the limitations inherent in the 3Ts and their fire control systems. It established high quality-control standards, created improved management and training school facilities that increased manning levels of personnel, and instituted an integrated maintenance plan to reduce system downtime. Once these actions were initiated, firing success rates improved and system availability increased for Terrier, Tartar, and Talos. This was an important achievement since these systems were the mainstay of the Navy's surface-to-air missile capability and would remain so until the as-yet-unnamed new system was developed.[3]

By the end of June 1963, Reich, in addition to his responsibility for fixing Terrier, Tartar, and Talos, had also acquired command over Typhon—the so-called fourth T—having taken over the small Typhon project office in the Bureau of Weapons (BuWeps) by mutual agreement with Rear Adm. Frederick L. Ashworth, assistant chief for research and development. Typhon was an integrated weapon system conceived by the Applied Physics Laboratory (APL) to defeat the air threats that were expected to emerge in the 1970s. The heart of the system was the multifunction, electronic scanning SPG-59 radar that would be capable of simultaneously tracking ten targets and guiding twenty missiles to intercept. The system would be capable of handling two different missile types and would be integrated with the Naval Tactical Data System/combat information center (NTDS/CIC). In the words of Alvin Eaton, APL's supervisor for the project, "Typhon was the first attempt to recognize that the anti-aircraft problem was a total package: that you had to consider finding potential targets, sorting them out, defining which the things you saw were proper targets for engagement, setting up an engagement process, and actually executing the engagements."[4]

Typhon was still a research and development (R&D) program far from production when Reich took over the twelve-man project office. He soon discovered that the proposed ship-building program for fiscal year 1963 included the construction of a nuclear-guided missile frigate (DLGN) equipped with the Typhon weapon system. This was problematic, because BuWeps would have to deliver a Typhon weapon system ready for installation a mere eighteen or twenty-four months after a development contract was issued. Even if the new system proved to be completely successful, Reich did not think it would be possible to deliver a deployable system in this time frame. Typhon was a long way from going into production, which he soon found out after visiting both Westinghouse and APL. After several months of briefings and visits to the plants that were putting the hardware together, Reich found that the only aspect of Typhon that was demonstrable was a very small laboratory model of the radar. While the model showed that under certain circumstances the phenomena upon which the system was being built worked—in theory—there was no evidence it could be perfected in any reasonable time frame. The Navy, Reich rightly concluded, could never build a Typhon ship under the 1963 fiscal year program.[5]

To discuss his concerns about the Typhon DLGN, Reich went to see Admiral Stroop. In his view, he told Stroop, BuWeps would be unable to supply the hardware necessary to equip the ship with the weapons system currently under development. Reich suggested that they cancel the DLGN and transfer the funds to support the Get Well program effort instead. Admiral Stroop, according to Reich, "understood and he agreed 100 percent, but he told me I had [no chance at all] of doing it," because the forces at work in OpNav wanted a new system. Stroop did not think that Reich had much of a chance of succeeding but gave him his blessings and told him to go ahead and try. Reich made some flag officers in the Office of the CNO very irate when he began to float the idea about canceling the '63 fiscal year Typhon frigate. He was able to convince Secretary of the Navy Korth however, who persuaded Secretary of Defense Robert McNamara to cancel the Typhon shipbuilding program in December 1963. The DLGN was dropped from the budget after $190 million had been appropriated. The sum of $121 million from the canceled DLGN program was then diverted to the Get Well program. Typhon itself was canceled shortly thereafter. It died largely because of the immense cost and complexity of its fire control system and the size of the SPG-59 radar, which made a Typhon ship prohibitively expensive.[6]

After it became clear during the final months of 1963 that the Typhon program would not be continued, Admiral Reich and Dr. Wakelin persuaded Chief of Naval Operations Adm. David L. McDonald to establish a new program to develop an advanced surface-to-air missile system that would enable new cruisers and destroyers to defeat massive, sophisticated Soviet air attacks in a heavy ECM (electronic countermeasures) environment—something even the upgraded 3Ts could not accomplish. Reich was told to go forward with the program and was "instructed to lay down a program to develop what would be a better technical approach" than that which had been used for Typhon. To this end he was appointed director of the Advanced Surface Missile System (ASMS) project established within the Office of Naval Material in December 1963.[7]

Cancellation of the Typhon program forced the Navy to reorient and consolidate work on its surface-to-air missile programs, focusing its resources on the twin goals of achieving the full operational potential of missile systems already in place while meeting the evolving changes in the air threat to the fleet caused by the rapid changes in technology. Because the 3Ts would remain the mainstay of the Navy's surface-to-air capability for the foreseeable future, it was essential to continue to improve these missiles. The Navy, according to the testimony given before Congress by Robert W. Morse, the Navy's assistant secretary for research and development, thought improvements could be best accomplished with a program to develop a new missile that had a high degree of commonality with, and that could serve as a replacement for, the current missiles. The result was the Standard Missile program, authorized on November 27, 1963.[8]

The goal of the program was to produce a "standard" missile that would replace the Terrier and Tartar missiles on a one-for-one basis. The name of the new missile, Standard Missile (SM-1), appears to have been chosen to appeal to Secretary of Defense Robert McNamara's penchant for commonality. Two versions of the new missile were developed: a medium-range version that would serve as an upgraded Tartar, designated the SM-1 MR; and an extended-range version that would serve as an upgraded Terrier, designated the SM-1 ER. Both missiles would be engineered to be fully compatible with existing weapons direction equipment, fire control, radar, and launching systems, and they would gradually supplant the old missiles in the Navy's stockpile.

As Reich took charge of the ASMS project, relations with the directorate of the Naval Ordnance Laboratory, Corona (NOLC; located in Corona, California) became strained when he began to question the objectivity of the Missile Evaluation Division that was charged with providing an accurate and impartial evaluation of the Navy's missile firings. Reich had first become involved with NOLC while he was in command of *Canberra*. At the time, he requested their assistance in evaluating the Terrier missile firing trials that he would be conducting during the ship's Mediterranean cruise. Dennis Casebier, then a brand-new employee of NOLC's missile evaluation department, and a colleague were dispatched to *Canberra*. After analyzing the firings of *Canberra*'s battery of RIM-2 Terrier missiles, they concluded that the weapons were indeed faulty. "We confirmed the captain's suspicions," said Casebier. "We told the truth." What he failed to realize was that many of the analysts employed by NOLC felt pressure to withhold negative feedback on the various missile systems they tested to preserve the reputation of NOLC's design, development, and delivery teams that were responsible for the missile's fusing system.[9]

Reich visited NOLC early on and had an opportunity to review their reports. By the time he was appointed head of the ASMS project, he had become skeptical about NOLC's reporting and had detected a conflict of interest. Reich's people had reason to believe that the fuses were not performing as well as they might, yet seldom if ever would there be any comment on fuse performance, and what comments did appear were always very complimentary. Reich realized that there was a real need to ensure objectivity in the manner in which the missile firings were being analyzed. "Somebody," he would later say, "had to raise the question, somebody had to recognize there was a problem, and somebody had to do something about it." That was in keeping with his persona. With the rank of a rear admiral and with his status as director of the ASMS, Reich had the wherewithal to do something about it.[10]

In the effort to resolve the problems associated with determining missile systems' performance and describing it in a consistent fashion, Reich recognized the need for a sound analytical model and database, and an unbiased, independent analysis agent who could use the model and the data. Reich also recognized that the NOLC Missile Evaluation Department had the technical expertise, models, and databases to perform the analysis, but that it lacked the direct reporting relationship required to provide the truly independent and unbiased reports the Navy needed. Under Reich's leadership, and with Admiral Stroop's help, the Missile Evaluation Department was separated from NOLC and established as the Fleet Missile System Analysis and Evaluation Group (FMSAEG). It became a separate command at the Corona site. The establishment was officially authorized by the Secretary of the Navy on February 24, 1964. The mission assigned to FMSAEG was "to provide the Navy Department, the Operating Forces, and appropriate organizations of the Shore Establishment with evaluation of performance, reliability, readiness, and effectiveness of missile weapon systems, subsystems and assemblies, and associated test equipment and checkout systems."[11]

Reich never liked it when the senior people in the Navy either did not know about or wanted to close their eyes to particular problems or defects. Serious problems existed, even if the senior people did not want to listen to problems or acknowledge that there were problems. He was well aware of the need for public relations and optics, but he abhorred the need to brag about some ship or weapon system. Reich also knew that you could not always be raising hell—you had to accentuate the positive. Nevertheless, he was not afraid to rock the boat when necessary to get things done.[12]

In late 1963, Reich was instructed to go forward with the definition of an ASMS based on the Temporary Specific Operational Requirement (TSOR) issued by Secretary of Defense McNamara in early January of that year. This was a document that described the technologies that could make the proposed system a reality and that would form the basis for the Navy's operational requirement. By this time the Navy had an alternative Outer Air Battle Program to counter the threat of Soviet bombers: a long-range air-to-air missile (later named the Phoenix) that would be carried by the Navy's version of the F-111 fighter, which was under development. Deployment of the F-111 would supposedly eliminate the need for a long-range missile (Talos) and the large ships needed to accommodate them. This would allow the Navy to develop smaller missile ships that were less costly and easier to maintain. The TSOR emphasized the need for the new system to defend against small, fast targets (i.e., guided missiles) closely spaced at low altitude, attacking under heavy jamming cover, and taking advantage of sea and land radar clutter.[13]

Reich's first task was to determine what a fully integrated missile weapon system should look like. To achieve this objective, he had the ASMS project office solicit proposals from contractors in the defense industry; each was to outline its concept for the system described in the TSOR. Fifteen contractors submitted initial proposals. Seven of these—Boeing, Hughes, RCA, Raytheon, Sperry, Westinghouse, and General Dynamics Pomona—were considered worthy of further study. In August 1964 contracts were issued to these companies for conceptual designs of the new system. The contractors were all asked to team up with subcontractors who had expertise in shipbuilding, electronics, or naval ordnance equipment. Each contractor was given a half-million-dollar, six-month study contract to review the state of missilry in the United States and to recommend how the Navy should proceed with the development of an advanced surface missile system. To ensure that all the teams would get a proper briefing on the status of the surface missile in the Navy, Reich saw to it that a contractor's school was established at the Naval Ordnance Laboratory in White Oak, Maryland; the school ran for about two weeks.[14]

To evaluate the proposals that would be submitted by the industry teams, Reich created the ASMS Assessment Group. It was staffed with 110 to 115 people recruited from the Office of Procurement, Acquisition, and Logistics; the ASMS project office; the Applied Physics Laboratory (APL); Bell Telephone Laboratory; the U.S. Army Missile Command; and the various contractors. To lead the group as its director, Reich asked Rear Adm. Frederick S. Withington, a former chief of the Bureau of Ordnance (BuOrd), to come back to active duty. Withington, an experienced technical manager, was well versed in the nuances of the 3T program, having been chief of BuOrd while much of the work on these systems had taken place. Reich tasked the ASMS Assessment Group with developing a baseline ASMS system configuration and assessing its effectiveness against the future air threats postulated by the Navy. The group, popularly known as the "Withington Committee," was assembled at an off-campus site in Silver Spring, Maryland, in January 1965. It continued to work until June. Subgroups were formed to consider operation requirements, to assess effectiveness, and to do top-level concept design for the ship system (less launcher), the launching system, and the missile. Ship installation, costs, reliability, and logistics support were also addressed.[15]

When the committee's report reached the Office of the Secretary of Defense in the late spring of 1965, the Army was proposing the creation of its own Surface-to-Air Missile Development (SAM-D) program to replace its Hawk and Nike-Hercules missiles. Because of the similarities of the two programs (the SAM-D system was much like a simplified Typhon system), Secretary of Defense McNamara directed the two services to explore ways to merge the programs into a common Army-Navy missile system. Although the study group established for this task determined that there was insufficient commonality for a joint project, McNamara remained unconvinced. He directed Dr. John S. Foster Jr., the director of defense research and engineering, to establish a joint project for commonality. When the contract definition for SAM-D was approved in March 1966, it included a proposal for a

Navy variant. This idea was scuttled in January 1967, after the SAM-D source selection advisory board, which had started holding its session the previous October (with several Navy officers in attendance) concluded that the carrying platforms and operating environments differed so greatly between the Army and Navy that a common system would cost more than separate systems for each service. It was only then that McNamara agreed to the development of two programs; and a directive was issued that the ASMS and SAM-D were to remain separate.[16]

By the summer of 1965, Reich had been immersed in the guided missile business for more than three and a half years. Since his selection to rear admiral, he had been sitting on the beach, and he knew that if he did not get sea duty as flag officer soon, his career would be over. It was true that he had a very important job, but he was not enhancing his career. His recommendation to cancel the Typhon DLGN that helped derail the Typhon project altogether did not win him many friends within the ordnance community. He had crossed swords with Chief of Naval Operations Adm. George W. Anderson and had a very cantankerous session with Deputy Chief of Naval Operations for Logistics Vice Adm. John Sylvester. Reich believed deeply in all his actions, but he was enough of a pragmatist to realize that he was not helping his favorability standing within the body politic of the Navy.[17]

Reich thought that Vice Adm. Benedict J. Semmes Jr., the current chief of the Bureau of Personnel, might help him. Semmes had been chief of staff to Commander, Destroyer Force, Atlantic Fleet (ComDesLant) when Reich was in command of Submarine Squadron 8. Reich had conducted numerous joint exercises with DesLant's ships during that tour of duty and had gotten to know Semmes fairly well. Reich went to see Semmes in July 1965 and told him of the discussion that he had had with Vice Adm. William R. Smedberg immediately after having been delegated to direct the Special Navy Task Force for Surface Missile Systems. Smedberg had been chief of the Bureau of Personnel at the time and had pulled him aside to inform Reich that he was going to stay put for a while. Smedberg was obviously aware of the importance of technical management continuity in complex new engineering projects. As Boslaugh has astutely noted, "Continuity of management was particularly important to keep a corporate history of the reasons for technical decision during the critical formative stage, and it also ensured that a change in managers could not bring about a disruptive change in system design." "I want you to know," Smedberg had said, "I don't want to hear anything from you in terms of change of duty for a minimum of three years, so you are not going anywhere." Semmes listened to Reich, then told him how important the 3Ts were, the great job he was doing, and the need to keep doing it. "I understand full well," replied Reich, "but you know as well as I do what the facts of life are in the flag officer business, and I've got to get to sea." Semmes said he would look into it.[18]

A few weeks went by before Semmes contacted Reich to tell him that he had a very good job for Reich. "We want you to relieve the commander of the Operational Test and Development Force." The Operational Test and Evaluation Force (OpTevFor), headquartered in Norfolk, was an independent agency within the Navy reporting directly to the CNO to oversee the operation testing and evaluation of new weapon systems, ships, aircraft, and missiles. As Reich later explained, "That went over like a lead balloon." Sitting on the beach in Norfolk was not going to sea. Reich told Semmes he would think about it and decided to take some action on his own. He went to see Vice Adm. Charles B. Martell, the assistant CNO and executive director of Antisubmarine Programs (OP-95). OP-95 was a new office within OpNav, established on February 17, 1964, to exercise centralized supervision and coordination of all antisubmarine planning, programming, and appraising to ensure an integrated and effective ASW effort. Reich told Admiral Martell that he wanted to go to sea with an ASW group. He agreed with Reich that OpTevFor was not a good place for Reich and that he was not opposed to what Reich wanted do.[19]

Reich went to see Vice Admiral Hayward, now commander of ASW Force, Pacific. As the reader will recall, Hayward had approved Reich's request to conduct test firings of the Terrier missiles while he was in command of the *Canberra.* Reich considered Hayward to be a good friend. He asked Hayward if he would accept him as one of his group commanders. Hayward said yes and advised Reich that he had already gotten approval from the commander-in-chief of the Pacific Fleet. After lining up support for his assignment as an ASW group commander in the Pacific Fleet, Reich went back to see Admiral Semmes. Semmes hemmed and hawed. He did not say yes, and he did not say no. After a couple of weeks went by Reich decided that if he did not take some kind of strong action, he was going to wind up in Norfolk in OpTevFor. So, he went to see Secretary of the Navy Paul H. Nitze. Reich had a very candid discussion with Nitze, telling him that he had served his time, wanted to go to sea, the Pacific Fleet was happy to have him, and there was a job available. Would he help, asked Reich. Reich did not think Nitze was happy about it, but Reich got command of ASW Group 5 in the Pacific Fleet because the Secretary of the Navy directed it. This confirmed Reich's belief in the old adage, "God helps them who help themselves." And that is how Reich left the 3Ts business in September 1965.[20]

» Eli Reich's 1935 *Lucky Bag* photograph

U.S. Naval Academy

» Reich's first ship, the USS *Pensacola* (CA 24), at sea, September 1935. Reich learned a lot while supervising the ship's R Division. *Naval History and Heritage Command*

» The World War I–designed destroyer USS *Waters* (DD 115) at anchor during the 1930s. *Waters* was the first of three four-stackers that Eli Reich would serve in between March 1936 and May 1938. *Naval History and Heritage Command*

» The *R-14* (SS 91), Reich's first submarine, was a twenty-year-old obsolete boat with flawed diesel engines. As the boat's engineering officer, Reich spent an inordinate amount of time making sure they kept running. *Naval History and Heritage Command*

» The wreckage of the *Sealion* (SS 195). It was destroyed by Japanese bombers while docked at the Cavite Navy Yard on December 8, 1941. Reich had served as the boat's engineering officer from the time he went on board in August 1939. *Public domain*

» The *Lapon* (SS 260), with Reich on board as executive officer, slides down the ways at the Electric Boat Company's yard in Groton, Connecticut, on October 27, 1942. *NARA*

» Cdr. Eli T. Reich, USN, commanding officer of the *Sealion* (SS 315). Reich would receive the Navy Cross with two Gold Stars for his actions in command of the *Sealion*, including the destruction of a Japanese battleship, the only one sunk by a U.S. submariner. *NARA*

» The *Fletcher*-class destroyer USS *Compton* (DD 705), shown here in her World War II camouflage, was Reich's first surface ship command. *NARA*

The Recommissioning
of the
U.S.S. Stoddard
DD - 566
at
Charleston Naval Shipyard
March 9, 1951

U. S. S. STODDARD (DD-566)

The U. S. S. STODDARD (DD-566) is a 2050-ton destroyer of the Fletcher class built by the Todd Pacific Shipyards, Inc., Tacoma Division, Seattle, Washington; her keel being laid 10 March 1943.

The U. S. S. S. STODDARD was named in honor of James Stoddard, a seaman aboard the U. S. S. MAMORA, who performed distinguished combat service during the Civil War. The U. S. S. STODDARD was launched 19 November 1943 and first commissioned 15 April 1944, at which time Mrs. Mildred Gould Holcomb, wife of Captain Harold R. Holcomb, USN, served as the official sponsor. During World War II the U. S. S. STODDARD earned three engagement stars on the Asiatic Pacific Area service ribbon for her part in the Kurile Islands operation, the assault and occupation of Okinawa Gunto and Third Fleet operations against the Empire Islands of Japan. She also earned the Navy Occupation Service Medal, Pacific for her operations during the period of September 2 to November 19, 1945. She was placed out of commission, in reserve, by directive dated January 1947.

The present activation of the U. S. S. STODDARD (DD-566) has been under the direction of Commander Charleston Group, Atlantic Reserve Fleet.

» Cover of USS *Stoddard* (DD 566) recommissioning pamphlet. Reich, the ship's commanding officer, was so upset with her condition that he refused to sign the acceptance papers. *NavSource*

» *Aucilla* (AO 56) underway with Reich in command. This was his first deep-draft ship. *NavSource*

» The guided missile cruiser *Canberra* (CAG 2) underway on January 9, 1961, with Reich in command. His experiences with the ship's Terrier missiles led to his assignment in charge of the Advanced Surface Missile System project that laid the foundations for Aegis. *U.S. Navy*

» Wayne Meyer's first ship, the destroyer *Goodrich* (DD 831). He was assigned duty as the ship's electronics officer due to his recently received degree in electrical engineering from MIT.

Naval History and Heritage Command

» The light cruiser *Springfield* (CL 66) was a much larger ship than the *Goodrich* (DD 831). Although Meyer was the ship's electronics officer, he learned a lot about the ship's combat information center. *NARA*

» When Meyer joined the destroyer tender *Sierra* (AD 18), he was the only one of three line officers on board—the captain, the executive officer, and himself. In addition to his duties as ASW/CIC officer, he also had to stand deck watches. Meyer had a very dissatisfying cruise and couldn't wait to get off the ship. *U.S. Navy*

» The radar picket destroyer *Strickland* (DER 333) taken a year after Meyer serviced with her in the Atlantic *Naval History and Heritage Command*

» A Talos missile being launched from the guided missile cruiser *Galveston* (CLG 3). Meyer claims that he fired more Talos missiles than anyone else while he was the ship's fire control and weapons officer. *Naval History and Heritage Command*

» Commissioning ceremony for the *Arleigh Burke*–class guided missile destroyer *Wayne E. Meyer* (DDG 108) *U.S. Navy photo by Mass Communication Specialist 1st Class Tiffini Jones Vanderwyst*

CHAPTER 18

ELI REICH'S LAST SEA DUTY

ON OCTOBER 27, 1965, REAR ADMIRAL REICH, at a ceremony conducted on board the anti-submarine carrier *Bennington* (CVS 20), relieved Rear Adm. Robert A. MacPherson and assumed command of Anti-Submarine Warfare Group 5. Shortly thereafter, Reich moved his flag from *Bennington*, which was scheduled for overhaul, to the anti-submarine carrier *Kearsarge* (CVS 33).

Having just returned from a deployment, the anti-submarine warfare (ASW) group stood down during November and December. Reich used these two months—as he had done after taking command of *Canberra*—to review the operations of his command during its previous deployment. It also gave Reich the opportunity to attend some of the senior courses at the Anti-Submarine School in San Diego.[1]

Looking over the command setup on the *Kearsarge*, he discovered that the group commander's battle station was on the flag bridge, directly under the navigation bridge. The flag bridge was very open; from the bridge you could observe the flight deck and see the horizon. This was fine, as Reich later pointed out,

> if you as a group commander were concerned with the immediate operation of the screen of the task force and your purpose was just the tactical command of carrier and screen. . . . But in order to coordinate, to command and control an ASW group, you are not limited in any sense by the horizon. You are limited really by the range of your sensors. . . . What is important is not what you can see visually from the flag bridge of the carrier but what your sensors can detect.[2]

Reich immediately recognized that this would cause a problem for him. Having just come from managing a major new weapon system, he was very cognizant of the rapid advances being made in sensors, fire control, and the integration of tactical data. It was obvious to him that a system engineering approach needed to be undertaken with regard to the overall objective of the

ASW group. Reich wanted to do something about it, but he realized that his tenure was limited—just one and a half years at most. He was aware, too, that ASW Group 5 was going to deploy again in six months, and the time he had in which to do something was very limited.[3]

A year earlier, he had given a briefing to Dr. Reuben R. Mettler, the president of TRW Systems Operating Group. Mettler was impressed with what Reich had accomplished with the 3Ts, and the two became quite friendly. Reich was aware that Mettler had spent a considerable amount of time looking at the Navy's approach to ASW for the Secretary of the Navy. He also knew that TRW was supplying systems engineering and test support for the Navy's ASW program. TRW's headquarters in Redondo Beach, California, was only seven or eight miles from *Kearsarge*'s mooring in Long Beach, so it was easy to arrange a meeting with Mettler to discuss his ASW command and control problem. Dr. Mettler immediately recognized this as an opportunity for TRW and offered to make a team of project engineers available. After obtaining approval from Admiral Martell and Admiral Hayward, Reich brought a group of TRW engineers on board *Kearsarge*. Their task was to immerse themselves in the minutiae of command and control in an ASW force. The technical group, which began with three or four engineers, at one point included eleven engineers.[4]

Within four or five months, Reich, working with the engineers from TRW, established a new command and control center for the ASW group commander in a space directly adjacent to the ship's combat information center (CIC). They also established an ASW intelligence center to house the Anti-submarine Center for Analysis (ASCA). The ASCA was where all the ASW data, or intelligence, came together. It was also where the ASW specialists interpreted the lofargrams* obtained from that *Kearsarge*'s S-2 Tracker antisubmarine aircraft.

In 1964, the Navy began to develop an expendable bathythermograph: a device carried by every submarine to measure the water temperature at different depths. Bathythermographs were important for submarine and ASW operations because they were used to locate the thermocline, which is the transition layer between the warmer water at the ocean's surface and the cooler water below. Operating below the thermocline makes sonar detection much

* A Low Frequency Analyzer Recorder that produced electromechanical displays on electrostatic recording paper that produced graphic images of sonar signals.

more difficult because of the various negative effects cold water has on the uniform transmission of sound waves. Destroyers equipped for ASW also carried bathythermographs on their fantails. These were cumbersome to use and difficult to deploy and retrieve. The advent of the expendable bathythermograph, known as the BT, would greatly facilitate the use of this device on surface ships and helicopters. Reich learned that the Navy had completed development of an expendable BT, but that they were not available to the operating forces. Exhibiting his relentless determination to see that his command had the best training and equipment, Reich, using a little arm twisting, was able to obtain a thousand of the new BTs for his ASW group.[5]

Reich's ASW Group 5 was made up of Destroyer Squadron 23 (DesRon 23), the *Kearsarge*, the flagship, and her air group. The air group included two squadrons of Grumman S-2 Tracker ASW aircraft, one squadron of Sikorsky SH-3 Sea King helicopters, and a detachment of five Grumman E-1B Tracer early warning aircraft. ASW Group 5 trained for deployment under Reich's command starting in January 1966 and continued through May.[6]

The E-1B Tracers carried long-range radar and communication gear and could spend extended time in the air keeping track of the S-2 Trackers, providing grid locations, and relaying radio communications. Reich was so impressed with the capabilities of the E-1B Tracers and their help in controlling the S-2 aircraft, that he spent quite a bit of airtime in them. The S-2 Trackers were the tactical units. They planted and monitored sonar buoys, and carried torpedoes and rockets to attack any enemy submarines they were able to locate.[7]

In mid-June 1966, Reich's ASW force conducted a graduation exercise known as the ORE, which stood for Operational and Readiness Evaluation, prior to deployment. It was a week-long undertaking conducted east of Pearl Harbor around Hawaii and Molokai. After completing the ORE, the group returned to Pearl Harbor where the results of the exercise were analyzed and critiqued.[8]

On the first day of July 1966, or thereabouts, Reich and the immediate members of his staff flew to Yokosuka, Japan, to prepare for the joint exercise with the Japanese Maritime Defense Force. Reich's ships departed Pearl Harbor for Yokosuka immediately after the Fourth of July. After a few days in Yokosuka, the group, which was now joined by the diesel submarine *Bonefish* (SS 582), left Tokyo Bay and sailed toward the Korean Peninsula. Once on station it began operating in alternating periods with the Japanese Maritime Defense Force and the Korean navy with a twenty-four-hour recess in between.

It was during one of these interim periods that an unexpected event occurred demonstrating that Reich had not lost any of the aggressiveness or tenacity that he had shown when in command of the *Sealion*.[9]

Bonefish, which ran submerged throughout the exercises, was off *Kearsarge*'s quarter, snorkeling at twelve knots with the destroyers screening ahead. It was a bright afternoon, and Reich was walking up and down the flight deck, as he liked to do for recreation. He was not thinking about anything in particular and looked around to see where *Bonefish* might be. He was also curious as to where the Russian destroyer was that had been shadowing them. When Reich saw the Russian off the starboard quarter, smoke was streaming out of the destroyer's stacks, and she was "up to speed." Reich wondered what was going on. After watching the Russian ship for a while, he realized that the destroyer was making a run on the *Bonefish* as an exercise.[10]

Reich did not like this one bit. He quickly went to the navigation bridge and asked *Kearsarge*'s commanding officer, Capt. Willard L. Nyburg, if he knew where the *Bonefish* and the Russian shadow were. When the captain and the other officers on the bridge saw what was happening, Reich ordered Nyburg to make an emergency turn and head for the *Bonefish* at the best speed *Kearsarge* could make, telling him, "We didn't bring the *Bonefish* into the Sea of Japan to operate as a target for the Russian navy." Next, he picked up the TBS (talk between ships) telephone on the bridge and called the *Bonefish*, stating his intentions to head for her and instructing the submarine's skipper to hold his position until he saw the *Kearsarge* coming and to then drop below the ship so that *Kearsarge* could run right over her. Reich then called the screen commander, repeated his intentions, and ordered the destroyers to follow the *Kearsarge*.[11]

When *Kearsarge* turned, heading for the *Bonefish*, the Russian destroyer was six or seven miles away. That began what Reich later termed a "game of chicken with the Russians." They were very brash, according to Reich's account, coming right into the task force, inside the screen at times. The Russian "saw the light" and moved out of the way. After the Russian moved off, the task force formed back up and the next morning picked up six Korean destroyers for another joint exercise.[12]

After finishing the exercises with the Korean navy, Reich brought the group into Sasebo, Japan, for a short respite. ASW Group 5 left Sasebo in mid-August. A few days later it arrived in Yankee Station in the Gulf of Tonkin, where it would spend the next five months providing surface and subsurface surveillance for the U.S. forces in the Tonkin Gulf and the north part of the China Sea. The surface/subsurface surveillance control (SSSC) mission required Reich's command to identify and track all the surface contacts in their area. They paid special attention to those ships that entered the gulf and proceeded to Haiphong Harbor, keeping them under surveillance and photographing them until they got dockside. "Sub-surface surveillance was a farce," in Reich's words, because there weren't any strange submarines in the Gulf of Tonkin. "We were in a passive mode as far as an ASW group and that's where our own submarines came in."[13]

While in the Gulf of Tonkin, Reich always had one or two submarines assigned to the group. Without the need to keep track of other submarines, Reich was able to organize a series of ASW training exercises for DesRon 23 and the other destroyers at Yankee Station. These were shallow water ASW torpedo exercises, conducted against their own submarines, that provided extensive performance testing of the Mark 44 ASW torpedo—a seawater-activated electric battery-powered acoustic homing torpedo weighing 433 pounds. The torpedoes, armed with exercise warheads, were programmed to not hit their targets. They would not damage the target or themselves and could be easily recovered. All the torpedoes, which were either dropped from S-2 aircraft or fired from the destroyers' deck tubes, were serviced in the *Kearsarge* torpedo shop. Reich claimed they fired close to 150 torpedoes. He also arranged for DesRon 23s destroyers to be routed on the destroyer gun line that was providing fire support to the troops fighting in South Vietnam.[14]

Reich's ASW Group 5 was also responsible for providing air-sea rescue with its SH-3 Sea King helicopters. This was a very important mission, given the air war being conducted over Vietnam and the number of U.S. aircraft that were being shot down. At least half of HS-6's (Helicopter Anti-Submarine Squadron 6) Sea Kings were designated for air-sea rescue. To accomplish this mission, they were fitted with additional communication gear, had an armor plate installed in the pilot area, had additional hoisting gear installed, and were armed with door-mounted machine guns. When U.S. forces were mounting a special attack, such as an intensive raid on Haiphong, Reich would make sure that some of *Kearsarge*'s air-sea rescue helicopters were on station to facilitate a recovery when needed.[15]

While Reich was in command of ASW Group 5, one of the most dangerous helicopter rescues of the Vietnam War was conducted by one of the Sea

Kings from HS-6. The event occurred on August 31, 1966, when Lt. Cdr. Thomas Tucker, the officer in charge of the photoreconnaissance detachment from Patrol Squadron 63 on board the aircraft carrier *Oriskany* (CV 34), was shot down while making a low-altitude photographic run on the shipping in Haiphong Harbor. Tucker ejected from the damaged RF-8 Crusader he was flying and parachuted into the secondary shipping channel of Haiphong Harbor. He landed in the water not far from a Soviet freighter that began to lower a lifeboat. Tucker's wing man, Lt. Cdr. Foster Teaque, made multiple strafing runs to prevent the Soviets and North Vietnamese motorized sampans in the vicinity from reaching Tucker.

While this action was taking pace, Teaque's frantic radio calls for rescue were picked up by the helicopter named *Indian Girl* piloted by HS-6's commanding officer, Cdr. Robert S. Vermilya, flying as the on-call rescue helicopter at the north Search and Rescue Station. Vermilya and his copilot, Ens. William E. Runyon, pushed the nose of the Sea King up and headed for Haiphong Harbor as fast as they could. Peter Fey, in his book about Air Wing 16 in Vietnam, describes what happened next:

> As their Sea King approached the mouth of the harbor, still 6 miles from the shipping channel, the crew looked up to see to SAMs streak overhead toward Teaque. AAA [anti-aircraft] machine guns, and mortar fire began zeroing in on the helicopter. At the same time, a flight of four VA-152 Skyraiders led by Lt. Cdr. William Smith and Lt. Jack Feldhaus heard Teaque's radio transmission; they gathered their wingmen and proceeded to Haiphong at maximum speed. . . . Orbiting over the harbor, Smith quickly spotted Tucker, who by this time had climbed into his raft in a frantic attempt to put some distance between himself and the ships in the harbor.
>
> That Tucker had landed in quite possibly the worst location possible did not deter Teaque's or Smith's flight of Skyraiders. The aircraft were in range of many coastal AAA batteries and ship's AAA. . . . Vermilya had been flying at 3,000 feet, supposedly below minimum altitude of the SAMs and still above the effective range of small arms. After watching two SAMs pass overhead, Vermilya quickly changed his mind, dropping to wave-top height. Big Mother, as the Large and ungainly Sea Kings were affectionately known, threaded her way through the harbor.

> Vermilya kept the helicopter so low that he believed some of the gun batteries on shore actually held their fire so as not to hit ships in harbor.[16]

As the Sea King neared Tucker, the Skyraiders began to strafe and bomb the guns along the shore. Vermilya, dodging shell splashes, shipping, and sampans, managed to fly right by Tucker. He hauled the helicopter around to the right and came over Tucker hovering as the downed airman grabbed the rescue sling. As Tucker was being hauled up, Vermilya yanked the Sea King back the way he came, flying the gauntlet through the shipping channel and the six miles of harbor before reaching the safety of the Tonkin Gulf.

For their actions in rescuing Tucker while under fire, Teague and Vermilya were each awarded the Silver Star, Runyon was awarded the Distinguished Flying Cross, and the two air crew were awarded Air Medals. The event was significant enough that Reich recalled it in detail, "I can't say enough for those chopper pilots. Their major mission was as an anti-submarine helicopter squadron, but here they were being used as part of the attack force and they were the recovery force."[17]

Reich's force departed the Tonkin Gulf in late November 1966 and retired to Yokosuka, where he transferred all operating procedures and briefed the replacement ASW group before sailing to Long Beach, California, where they arrived just before Christmas. Reich was detached from command and relieved in the last week in January 1967. He received temporary orders to report to the chief of the Bureau of Personnel in Washington, D.C., for later assignment. At the time Reich had reason to believe that he was under consideration to relieve Vice Admiral Martell, who was finishing his tour as director of anti-submarine programs in OpNav.[18]

CHAPTER 19

ELI REICH, ACCOUNTING GURU

WHILE REICH WAS COMING UP THROUGH THE RANKS he observed, and came to believe, that one of the shortcomings of the naval profession was its lack of consideration for logistics planning. It lacks glamour, he opined, and consists of "a myriad of detail." So, when he reported for duty in Washington, D.C, Reich was somewhat disappointed when Admiral Semmes, the chief of naval personnel, told him that he was being assigned to the logistics branch in the Office of the Chief of Naval Operations (OpNav). In late February 1967, Reich assumed duties as director of the Logistics Plans Division; then, in June, he became assistant deputy chief of naval operations for logistics.[1]

Toward the end of August 1967, Admiral Semmes told Reich that he was going to be moved to the Office of the Comptroller of the Navy to relieve the deputy comptroller, Rear Adm. Paul Masterton. Reich still had hopes of relieving Admiral Martell based on discussions that he had been having with Rear Adm. Turner E. Caldwell, one of Reich's Annapolis classmates. At one time Caldwell had been deputy to Martell, and Caldwell had been promised he would one day relieve the commander of the Atlantic ASW force, Rear Adm. Charles E. Weakley. It was not to be. Masterton went down to Norfolk to relieve Weakley, Caldwell came to Washington to relieve Martell, and in October 1967, Reich relieved Masterton, becoming deputy comptroller of the Navy.[2]

The situation in the Office of the Comptroller when Reich became deputy comtroller was highly unusual. Charles F. Baird, then–assistant secretary of the Navy for financial management, had moved up to become under secretary of the Navy for financial management and comptroller. By statute, the assistant secretary of the Navy for financial management was also the comptroller of the Navy. Until a replacement was found, the deputy comptroller—Reich, in this case—automatically became the acting comptroller for the Navy. This left Reich as the senior man in the comptroller's office. Reich operated this way until the first of the year, when Charles Bowsher, who was a partner in the Arthur Anderson Company, was confirmed as assistant secretary of the Navy for financial management.[3]

In June 1968, Bowsher asked Reich to look into press reports concerning the losses being written off by some of the Navy's most important shipbuilding firms. Rather than examining a single contact, Reich asked the team he established for the study to focus on all the major contracts held by specific shipbuilders. The study, which was finished by the end of September, was ten to twelve pages long with another ten or twelve pages of exhibits. The report identified at least $600 million of potential claims against the Navy's shipbuilding program. Privately Reich told Bowsher that it might even exceed a billion dollars. The most significant thing that came out of the study was Bowsher's recognition that the Navy had a big financial problem that was much greater than that which, according to Reich, had been reported in the *Wall Street Journal*.[4]

Shortly after the shipbuilding study was issued, Bowsher called Reich into his office and said, "You know, I'm concerned also about some of these major acquisition programs we have and how much we here in the comptroller's office know about the potential for problems. Of course, we're talking of financial problems. I think you ought to look at the new budget that's being planned. We ought to look at items of significant dollar value and then—we can't do them all—we ought to select one and do a little study in depth as to not only pricing out the requirements for the budget under study, but also . . . in the whole projected life of the project." Bowsher told Reich that he needed to form a special group to work on the project.[5]

Reich began by flagging certain high-value line items in the budget to see what funds had gone into each in prior years and to find out more about them. Reich assigned anywhere from two to five people from the special project group he had established to review a particular item. He also called in the services of outside experts as needed. A typical program review conducted in this manner was undertaken for the Mark 48 torpedo: a long-range, high-speed, submarine-launched torpedo being developed by Westinghouse to replace the Mark 37. After reviewing the Westinghouse program, it was obvious to Reich that the Mark 48 program had serious financial problems. Each Mark 48 torpedo, according to Reich's review, was going to cost three-quarters of a million dollars. This came as quite a shock to Reich considering that the cost of the Mark 37 in 1958 was only $38,000.[6]

In addition to the two auditing tasks Bowsher had requested that summer, Reich also received a one-page letter from Chief of Naval Operations Adm. Thomas H. Moorer, charging Reich with personally conducting an investigation into the activities of the Naval Audit Service in connection with a report on the expenses related to aircraft modification kits. The report, which was highly detrimental to the naval aviation community, had been provided to the Office of the Secretary of Defense. What Admiral Moorer was saying, in effect, according to Reich's interpretation, was "that the Naval Audit Service was a viper in our bosom and had been raising hell in the matter of investigating and reporting with their internal audit reports." The tone of the note, which was personal and hand-delivered, indicated that along with the audit report in question, the CNO was unhappy with the comptroller's office and with Reich personally. Admiral Moore was dyed-in-the-wool naval aviator* and would obviously take affront to anything besmirching naval aviation.[7]

Reich quickly determined that the report in question had been prepared eighteen months earlier in the San Diego Regional Office of the Auditor General. With the assistance of Capt. Eugene H. Auerback, the auditor general, Reich scheduled a meeting and flew out to San Diego to hear firsthand what they had to say. To his great surprise, they stood by their report documenting the tremendous number of spare parts and modification kits in warehouses on the West Coast. They told Reich that this had been going on for ten or twelve years, even though they had reported it many times. The particular report that Reich was looking into, they explained, was just another one of the numerous similar reports they had written in the past.[8]

During the next few days, Reich toured the warehouses in North Island. He found that they were, in fact, loaded to the gills with packing boxes and crates piled forty to fifty feet high, box on top of box for as far as one could see. While inspecting one of the warehouses, Reich pointed out one of the biggest boxes and asked the warehouseman who was accompanying his retinue to find out what was in it. The man pulled out the bill of lading and said, "This is a fuel cell for a helicopter." It was dated 1956 and had been sitting there for twelve years.

"What's the unit value?" Reich asked.

"$20,000."

"What's it used for?"

---

* No. 4255. He received his wings on June 12, 1936.

"That used to be used in a helicopter, but it's no longer usable because the rubber has deteriorated, and we no longer have that model."

"Why do you still have it here?" Reich asked.

Their warehouse man did not know.[9]

No one was working in most of the warehouses Reich entered except for one warehouse in which Reich discovered a group of fifteen or twenty men in the process of opening crates. Inside one of the crates were a large number of little packages that were being taken out of the excelsior packing in the crate. Reich was curious to know what was going on. It was the middle of the war in Vietnam, and it turned out that the men were looking for a particular gyro that was used in the F-4 Phantom's fire control system. There were not any in the supply system, he was told, and they were looking for spare gyros that they suspected were in this particular crate.[10]

When Reich reported his findings to Admiral Moorer, there was no question in Reich's mind that the admiral's "dander was up, and he was unhappy." He ordered that Reich provide him with a written report on his findings on the problematic audit every month, and that he explain in person what was going on. It was obvious to Reich that the head of the Naval Air System Command had told the CNO that the Naval Audit Service was giving ammunition to the Office of the Secretary of Defense and that the command was having one hell of a time getting adequate funding for the spares they needed. This was an ongoing problem for Naval Air System Command because Moorer could not force the Naval Audit Service to discontinue their reports.[11]

In the spring of 1970, Reich was rounding out his second year as deputy comptroller. As he later explained, "I was getting a little bit itchy as to what was forecasted for my future. I was unhappy about losing out on the opportunity to relieve Martell and, of course, regretted not having the third star. My interplay with the CNO, Admiral Moorer, in connection with the audit problem further added to my misgivings about what future, if any, the Navy had for me." Reich felt that he had been given a consolation prize when he was assigned as deputy comptroller, and he believed that Masterson had been elevated to three stars because he had been classmates with the CNO. And while Reich considered Masterson to be "a damned good attack aviator," he "was not an ASW fellow . . . it wasn't a question of what the billet was—the question was three stars, and so the powers that be decided who was going to get three stars." The decision was not based on who was better suited to be the head man in the ASW billet.

So, Reich was feeling pretty glum. He felt that the Navy had given him as much as they were going to. Bowsher was sympathetic but had no way of helping Reich until he received a call from Assistant Secretary of the Navy for Installations and Logistics (I&L) Harry J. Shillito. Shillito, Bowsher learned, was interested in making changes to the I&L office and was going to establish a three-star billet in connection with material acquisition and production engineering. Could Bowsher recommend someone? Bowsher immediately thought of Reich.[12]

"Would you think about it?" he asked Reich.

"I certainly would," he responded, knowing that he was well qualified to handle matters in weapons procurement and production engineering. And that was how Reich became deputy assistant secretary of defense for material[†] in June 1970. He would not have been nominated had it not been for the strong support he received from Bowsher.[13]

Reich's immediate job was to implement the reorganizations that Shillito wanted to carry out. These were in accordance with the major policy changes that Secretary of Defense Melvin Laird, Deputy Secretary of Defense David Packard, and Director of Defense Research and Engineering Dr. John S. Foster Jr. had laid out. Improving the acquisition policy for large weapon systems and major ships was one of their major concerns. Problems with the acquisition of major systems (overruns and inefficient contractor performance, for example) in the previous administration under Secretary of Defense McNamara had led to negative publicity in the press, congressional investigations, and Government Accounting Office reports. The new administration responded by making changes in the Department of Defense (DOD) acquisition policy.[14]

In May 1969, Deputy Secretary of Defense Packard formed the Defense Systems Acquisition Review Council (DSARC) to serve as an advisory board to the secretary of defense on matters concerning major weapon systems. A year later, one month before Reich became deputy assistant secretary of defense for material, Packard had issued another policy memorandum on defense acquisition, articulating the broad themes that would become the basis for the Department of Defense's new directive for procuring large weapon systems. Within a matter

† His title changed to deputy assistant secretary of defense for production and material on November 3, 1970.

of weeks after he took office, Reich was a member of the working group writing the new directive and was the principal contact in Shillito's office for the directive and its implementation.[15]

The first directive that Reich helped to write, DOD Directive 5000.1, was issued one year later in July 1971. It was only seven pages long and described the duties of the Secretary of Defense, the director of defense and engineering, and the assistant secretary of telecommunications. Although it applied to all acquisition programs, it referred specifically to major programs designated by the Secretary of Defense that had an estimated research, development, testing, and evaluation costs of more than $50 million ($373 million today), or an estimated production cost in excess of $200 million ($1.5 billion today). In addition to the duties of the Secretary of Defense and the two other principals of the DSARC committee, DOD 5000.1 laid out the three major decision points that had to be followed for all major procurement programs. These were DSARC I, program initiation; DSARC II, full-scale development; and DSARC III, production for deployment.

For a major project to be initiated, it had to have DOD approval and funding before it could begin the research phase under the DSARC I. When a project funded under DSARC I appeared to show promise, a decision had to be made to determine whether it should go into engineering development stage, that is, from the breadboard stage to preproduction under DSARC II. The decision to go into full-scale production and subsequent deployment was made in DSARC III. DOD 5000.1 stated that the director of defense research and engineering would chair the DSARC I and II decision-making meetings. The assistant secretary for I&L (Shillito) would chair DSARC III, which was perhaps the most important of the three meetings since it would determine whether a project went into production. Shillito assigned responsibility to chair the DSARC III committee to Reich, who was to act as his stand-in. Thus, Reich presided over a committee consisting of the director of defense research and engineering, the comptroller of the Defense Department, the assistant secretary for systems analysis, and associate members representing the joint chiefs. These high-level offices determined the fate of every major procurement program in the Navy.[16]

In the three years that Reich worked for Shillito, the shipbuilding program was one of the major concerns under review. "It was nothing but grief," according to Reich, who had to deal with the problems at Litton Industries' Ingalls shipyard in Pascagoula, Mississippi. Reich became personally involved with it after Shillito visited the yard in September 1970. Shillito was concerned with what he had seen there and asked Reich to go to Pascagoula to investigate what was taking place.[17]

When Reich visited the yard in November 1970, Litton Industries had contracts to build five amphibious assault ships designated as LHAs and thirty DD 963 *Spruance*-class destroyers. Both ships were highly complex vessels incorporating new features. The LHA was a new class of amphibious assault ship that would also serve as a command ship. In addition to features such as a morgue, a hospital, and a dental center, it would incorporate a command center loaded with all sorts of complex electronics gear for communications and navigation. The contract required Ingalls to design, develop, and construct the LHAs in the new west bank shipyard at Pascagoula using the latest modern techniques in shipbuilding. Reich was impressed with the fact that the west bank of the yard was laid out like a modern shipyard that could easily accommodate a tanker or other commercial vessel. But he wondered about Litton's ability to build a complex warship. He was greatly concerned with the company's labor policies, which he called "a losing game in their recruitment of a working force." When Reich returned from Pascagoula, he told Shillito that he too shared his concerns and believed that "Litton was in deep trouble." A month earlier, Ingalls had filed a formal complaint over increased costs attributed to the Navy's cancellation of four of the nine LHAs that were eventually ordered by the Navy. The two parties agreed to restructure the contract with a new target price and a revised schedule. Ingalls submitted a target price of $1.039 million including a claim for $246.6 million. The latter was rejected by the Navy, leading to years of litigation.[18]

In November 1972, Reich attended the third summit conference called by Secretary of the Navy John Warner to discuss the ongoing problems with Litton as Shillito's stand in. The full-day meeting was held at the Culver City headquarters of Litton in Los Angeles, California. Reich was very uncomfortable throughout the meeting as he later explained, "I felt a little bit at sixes and sevens at that meeting because—I think I know something about contracting and about negotiating—I was really uneasy as I was sure that on the Litton side there were some pretty sharp businessmen and there was a lawyer or two among them. I wasn't impressed with the government side and consciously made a decision when I saw this meeting shaping up—I decided I wasn't going to say word one. I was leery of the manner in which this thing had been set up and I came away from the meeting disturbed."[19]

Reich took copious notes that he wrote up when he returned to Washington, which he then showed to Shillito, telling him that he thought that there was a lack of professionalism on the government's side. Shillito told him to discuss it with Secretary of the Navy Warner. Reich went to Warner to tell him what he had discovered. Warner said, "Fine. Just leave it [the report] with me." And that was the last the Reich heard about it.[20]

Reich made one more trip to Pascagoula with Dr. Foster during the summer of 1973. Not much had changed. When he got back to Washington, Foster reported that he had serious doubts as to Litton's ability to carry out the contracts and again brought up the difficulties in recruiting a work force, the fact that they had people of questionable ability in positions of responsibility, and doubts regarding Litton's shipbuilding know-how.[21]

The Navy would continue to have problems with Litton for the next six years, until June 1978, when a settlement was reached that awarded Litton $447 million ($1.823 billion today). By then Reich was long gone, having retired on November 1, 1973, at the age of sixty, after thirty-eight years of active duty.

CHAPTER 20

WAYNE MEYER HONES THE SKILLS OF A MISSILEER

WHEN MEYER TOOK OVER THE TERRIER FIRE CONTROL DESK there were three different fire control radar systems in the fleet (see table 20.1). This situation was partly the result of the past practice of spreading missile ship conversions and upgrades over a series of individual ship overhauls. In 1964, a decision was made to accomplish all the pending improvements in a single program. Thus, the DLG Anti-Air Modernization Program was initiated. The objective of the program, which encompassed the nine ships of the DLG 16 class as well as the ten of the older DLG 6 class, was to improve the levels of anti-aircraft warfare effectiveness by incorporating major improvements in missile and associated weapons control systems. The program sought to standardize the Terrier weapons systems on all DLG classes, which would benefit logistics support, ease maintenance, and facilitate the assignability of personnel. All ships would be equipped with an improved three-dimensional air search radar, an improved guided missile fire control system, and the Naval

TABLE 20.1

TERRIER FIRE-CONTROL RADARS (circa 1964)

RADAR SET	DESCRIPTION	SHIP INSTALLATIONS
SPQ-5	3-axis tracking antenna	DLG 6, 7, 14, 5; CAG 1, 2
SPQ-5A	Beam riding Terrier (BT) only variant	DLG 8–13, DLGN 25, CGN 9
SPQ-5B	Upgrade to illuminate and guide Terrier (HT) as well as retain BT capability	DLG 16, 17, 19, 21, 22

Sources: William W. Berry, "Radars and Computers aboard USS *Sterett* and Other Warships during the Vietnam War," *NavWeaps*, accessed September 30, 2022, http://www.navweaps.com/index_tech/tech-088.php; Norman Friedman, *Naval Radar* (Annapolis, MD: Naval Institute Press, 1981), 181.

Tactical Data System (NTDS). The scope of the modernization involved the replacement of the primary three-dimensional SPS-39 radar with the SPS-48, the installation of the NTDS, the updating or replacement of missile systems to attain a uniform SPG-55B and Mark 119 computer fire control configuration, and the modification of the existing Mark 10 guided missile launching systems for compatibility with the Standard Missile.[1]

The name, "Naval Technical Data System," had been coined by Cdr. Irvin McNally in the fifteen-page concept report he prepared in 1955. As described by Boslaugh:

> McNally envisioned new shipboard radars designed specifically to work with the system. First, there would be a new two-dimensional long-range air search radar with a 300-mile aircraft detection range, and there would also be an accurate height finding radar. . . . Airborne early warning (AEW) radars carried aloft by specially outfitted aircraft were already operating in the fleet and were capable of sending real-time analog radar picture down to shipboard radar repeaters.
>
> . . . McNally called for each shipboard unit to handle data on 1,000 tracks of all types—hostile, friendly, or unknown; surface ship, air, or submarine; and derived from any sensor such as radar, IFF receivers, sonar, electronic countermeasures receivers, or other inputs. . . . The system was to compute the course and speed of each target and augment the raw video PPI [plan position indicator] display with electronically generated symbols and vector lines representing the location of each target and its course and speed.
>
> . . . McNally also specified that, on operator demand, the system should numerically display course, speed, altitude, and other amplifying information on a selected target.
>
> Once target position data was in the system, and courses, speeds, and amplifying information had been computed, or entered, for each target, McNally called for the ability to automatically assess all hostile and unidentified air targets to determine those most threatening to the task force. Next, he called for the computer to recommend the most appropriate interceptor, missile system, or AA guns to be assigned to the incoming target, and to graphically show weapon/target pairing and engagement status. In the case interceptor assignment, the system

> was to automatically generate interceptor speed, heading, and altitude directions for target intercept. He did not see this proposed automation as a way to reduce the number of men in the CIC, but rather as a way to allow the users to be more effective by concentrating on decision-making rather than repetitive manual actions.[2]

When Meyer joined the SMS project, the first production version of the NTDS was still being assembled while the *Chicago* (CA 136) was being converted into a Terrier/Talos guided missile cruiser. He would have his hands full integrating the NTDS into the Terrier fire control system.[3]

One morning in July 1964, Capt. Roger E. Spreen, the director of the Terrier program in Reich's Surface Missile Systems (SMS) project, sent for Commander Meyer and his assistant, Lt. Cdr. Edward J. Otth. After musing about the invention of the transistor and the Navy's failure to embrace its technological and logistic advantages, Spreen instructed them to come up with a plan to lead the development effort to "'transform' the Terrier In-Service Fleet to transistor-based designs, thereby fundamentally displacing the vacuum tube *wherever application permitted* [*sic*]." He told them, "Build a little, test a little, learn a lot"—a maxim that Meyer would adopt as gospel. Spreen also said, "I'll raise the money, and I'll keep the old man [Reich] off your back; I want results out of you." They went to work and spent the next year and a half getting everything ready for the *Leahy* (DLG 16), the first DLG scheduled to be modernized.[4]

The *Leahy*, which was commissioned on August 4, 1962, had a standard displacement of 4,650 tons, was 533 feet in length, and had a 55-foot beam. She was equipped with an SPS-39 three-dimensional search radar, two twin Terrier launchers, four SPS-55A fire control radars, and a Mark 76 mod 0 fire control system that was capable of handling both the BT and HT versions of the Terrier missile (SAM-N-7 BT-3 and SAM-M-7 HT-3). The ship, according to Meyer, had dozens of file-sized electronic cabinets loaded with "thousands" of vacuum tubes that Captain Spreen wanted replaced. Not all the vacuum tube triodes and diodes could be replaced by transistors at this time due to limitations in bandwidth and power requirements. Nevertheless, the Sperry Gyroscope Company, which designed and built the SPS-55 using transistors to replace many of the bulky vacuum tubes, was able to reduce the number of electronic cabinets from the six that were required for the SPS-55A to three in the SPS-55B. The SPS-55B, with 456 transistors, still had 900 vacuum tubes and 1,700 diodes.[5]

Replacing all the fire control equipment on the *Leahy*, including the four SPS-55 radars, was going to be a nightmare. Meyer, remembering the extended time it took to convert the *Galveston*, had an idea. It took an extraordinary amount of time to complete the wiring on *Galveston*'s fire control switchboard. Due to the limited working space, only two people could work on the switchboard at one time. What we are going to do, he told Spreen, is remove all the fire control gear and take it to Sperry, in Great Neck, Long Island, on trailers. "And then we're going to modernize and rewire and rebuild the gear with your beloved transistors. And then we're going to put connectors on it all, and we're going to bring it back to the ship so that we can rapidly connect it back there." Up to that time, according to Meyer, no fire control gear had connectors on board a ship. To win acceptance of this concept, Meyer spent time in Long Island with Sperry, in Philadelphia at the Navy yard that would be doing the modernization work, and with various contractors and suppliers trying to get everybody to accept the technique of using connectors.[6]

In 1962, the NTDS and SMS project offices arranged to integrate the prototype weapons director system (WDS) Mark 11 and the second prototype SPS-48 radar with the Terrier and Tartar fire control systems installed at the NTDS training school at the Mare Island Navy Yard. Lt. Cdr. David L. Boslaugh, an engineering duty officer then assigned as an assistant NTDS project officer, was given the task of overseeing the development of the automated tests using the unit computer as the main tool. Univac was contracted to write the requisite computer programs, and once written, the programs were to be validated at the Mare Island test site. "The first problem in developing the testing programs," writes Boslaugh, "was with our own NTDS equipment. The keyset central unit had inputs for only eight two-speed synchro inputs, and four were already in use to provide ships' roll, pitch, heading and speed inputs to the unit computer. We wanted to also input range, bearing, and elevation from both fire control radars at the same time, a total of six more inputs." After reviewing the technical manuals with the help of the engineers from the Ford Instrument Company, which was responsible for building the fire control system, Boslaugh and the other engineers working on the NTDS decided to solve the problem by installing a computer multiplexing switch to the Mare Island missile fire control switchboard.[7]

In order to modify the fire control switches that were to be installed in the nine DLGs to be modernized, Boslaugh needed to obtain permission from Meyer, who was considered the "owner" of the switchboards. Mayer had a lot on his mind one morning when he received Boslaugh and William C. O'Sullivan, the civilian engineer responsible for the NTDS installation. Meyer, according to the account given by Boslaugh, "listened rather impatiently. But his interest picked up when we described how an NTDS system, by switching, could be formed into two separate smaller systems, each using one of the unit computers, and while one system supported normal CIC ops, the other system, with appropriate communications, could be used to test a missile system." Meyer was a strong advocate of running a daily system operating test (DSOT) of the missile system; but it was a manually intensive, time-consuming affair, requiring hundreds of operator steps, measurement taking, and the writing down of results. Using the second computer for testing would greatly reduce the time needed to conduct the DSOT. Adding the switchboard advocated by Boslaugh would enable the DOST to be automated. Meyer agreed to the switchboard change if Boslaugh and O'Sullivan could demonstrate that the proposed computer program could obtain an accurate measurement of radar alignment errors after one circuit of a test airplane.

Boslaugh went back to Mare Island and supervised the installation of the computer-controlled multiplexing switch and the associated wiring. By the time the test plane showed up, the programmers had loaded the radar alignment test into the computer. To conduct the test, the operators locked both fire control radars onto the test plane and began entering position and height inputs from the SPS-48 radar. After one circuit, Boslaugh reported the following:

> We could see that we had gotten a good set of measurements, and released the Navy Skyraider to go back home. The pilot answered back, "Is that all? I usually spend hours boring holes in the sky for you guys and this only took 15 minutes." Two days later I showed Cdr. Meyer the test printouts. He asked, "How long did this take? "Fifteen minutes for one circuit," I replied. He responded that he would support changing the fire control switchboards in the nine frigates, but we would also have to convince the Ships Change Review Board because all the ships were now on the building ways.

This turned out easier to accomplish than Boslaugh had imaged. While explaining the technical nuances before the ship's Change Review Board, the chair told Boslaugh that he did not know what the hell he was talking about. But if the SMS project office (Meyer) supported it, then it was OK.[8]

Meyer's plan to have all the fire control equipment assembled and checked out and only then shipped to the yard doing the modernization for subsequent installation using the cable connectors was instituted for the first time on the *Leahy*. This procedure allowed various subsystems to be checked out prior to shipboard installation. This permitted the necessary structural modifications and cable runs to be conducted concurrently with the assembly and with the check of the weapons system's components on shore, saving time and improving the quality of the installation. "The shipboard 'marriage tests' conducted after installation went rapidly," according to one captain, "with only minor discrepancies."[9]

When the *Leahy* entered the Philadelphia Naval Shipyard on January 27, 1967, Meyer had been transferred to the Naval Ship Missile Systems Engineering Station in Port Hueneme, California; it was the organization Eli Reich had established to supervise the Ship Qualification Trials. Meyer was appointed director of the Engineering Directorate and acting executive officer, with instructions from Reich's relief to clean up the mess left by the outgoing director and to get the contractors out and to bring in civil service personnel. To expedite the creation of the station, Reich had used his influence within the industry to staff the unit with engineers recruited from the contractors. According to Meyer, Reich "was going to create civil service billets out of whole air." Well, that never happened. Meyer got rid of them by mandate so they could be replaced by civil service folks who were cheaper and worked directly for the government.[10]

Meyer went back to the Philadelphia Naval Shipyard to be present on the day they were to install the new fire control system in *Leahy*. Everything was ready, the spaces were all prepared, and they had done a walk-through. Because they could not be 100 percent sure of every cable length, half the cables had connections on only one end. Later on, as they gained more experience, they would add connectors to both cable ends. Thousands, according to Meyer's account of the installation, came down to watch the lowboys* delivering the

* A lowboy is a specialized trailer commonly used to haul large equipment. It can carry oversized loads because its cargo bed is low to the ground.

various components making up the fire control system that had been pretested at land sites established for this purpose. It was breathtaking to Meyer because the switchboard did not have to be wired, rewired, checked, and rechecked. And it worked.[11]

The other DLG ships in the program were modernized in a "turnaround" program that was developed to upgrade the Terrier missile system. The major elements to be updated were removed from the ship immediately after the ship's arrival in the shipyard and sent back to the manufacturer to be modernized for one of the follow-on ships. The donor ship, in return, received an updated suite from one of its predecessors.[12]

PART IV

AEGIS AND AFTER

CHAPTER 21

GENESIS OF AEGIS

WHILE WAYNE MEYER WAS FOCUSED on the DLG modernization program, efforts to develop a new anti-aircraft guided missile system under the Advanced Surface Missile System (ASMS) project directorate began to move forward. Secretary of Defense McNamara's desire for a common missile system delayed the development of the ASMS for several years. It was not until September 7, 1967, two years after the start date envisaged by the Withington Committee that Chief of Naval Operations Adm. Thomas H. Moorer approved the specific operational requirements for the ASMS recommended by the committee. The operational requirements recommended by the committee laid the groundwork for the weapon system that would be named Aegis.[1]

Before further action could be taken to procure the system proposed by the Withington Committee, however, the operational requirements had to be reviewed by the Systems Analysis Office in the Department of Defense in accordance with procurement rules for new systems as established by Dr. John S. Foster Jr., the director of defense research and engineering. The vehicle to accomplish this was known as a development concept paper (DCP). A tentative DCP for the ASMS project was issued as part of the Navy's request for authority to enter into a contract, and it was submitted to the Office of the Secretary of Defense on October 2, 1967.

The DCP laid out the basic functions and performance requirements of the ASMS. The three basic functions spelled out were as follows:

Detection—finding the target;

Engagement—killing the target; and

Control—of the first two.

The five performance requirements, later called the cornerstones of Aegis, were as follows:

Fast Reaction Time—to cope with fast, low-altitude threats;

High Firepower—to survive a saturation attack;

Electronic Countermeasures and Clutter Resistance—to perform in severe natural and manmade environments;

High Availability—to be ready immediately when needed; and

Area Coverage—to provide an effective area defense for the task force.

The ASMS DCP also called for the development of a new missile with advanced guidance and an improved warhead. The ASMS advocated by the Withington Committee was finally authorized on April 29, 1968, when DCP-16 was signed by McNamara's successor, Secretary of Defense Clark M. Clifford.[2]

The Navy did not begin the acquisition process until October 1968, when it chose three of the original contract teams led by Boeing, General Dynamics Pomona, and RCA to conduct contract definition studies for the ASMS weapon system. The purpose of these studies was to provide the contractors with the opportunity to propose a system that would meet the specified performance requirements and define the costs for a full-scale development program to produce the system specified by the Navy. Along with the system performance specifications, the Navy imposed a number of other requirements that had to be met. Included were the following:

- The ASMS was intended to go aboard DLGN 38 *Virginia*-class nuclear-powered destroyers and should be designed accordingly.
- The system was to use the Mk-26 guided missile launching system then being developed.
- Command and control would be provided by the NTDS installed in the ship.
- The ASMS equipment was to be contained in two modular deckhouses located fore and aft in the ship.
- The system was to use the latest Navy standard computers and displays.[3]

In the spring of 1969, the Navy assembled a Source Selection Evaluation Board to evaluate the reports submitted by three contractors. The board, which was composed of engineers and experts in management and finance, met in an abandoned facility at the Naval Support Facility Anacostia, in Washington, D.C. After several weeks of reviewing the proposals, the Source Selection Evaluation Board concluded that the specifications provided by each of the teams was insufficient and would have to be rewritten. The Navy provided some additional funds for this work and gave the contractors ninety days to rewrite the specifications and recost the program. The board was reconvened in September to evaluate the upgraded reports and conduct final technical negotiations with the three contractors, which continued through the first three weeks of November.

The new missile that was to be included in the ASMS development program would use inertial navigation for midcourse guidance. The idea of using inertial navigation to control the missile's flight was the brainchild of Capt. Richard W. Anderson, the director of the Guided Missile Division within the Naval Ordnance Systems Command.* Anderson had learned about inertial navigation while studying for his master's degree in aeronautical engineering at the special program for military officers organized by Charles Stark Draper in his MIT Instrumentation Laboratory. Draper's laboratory specialized in inertial navigation and guidance and was developing the inertial guidance system for Polaris during Anderson's tenure. While working on the ASMS project, Anderson realized that creating an inertial navigation system and a radar data link that would fit into the 13.5-inch airframe of the ASMS missile would enable the missile to be guided to the immediate vicinity of the target so that the ship's illuminators would be needed only for the terminal guidance phase (by contrast, illumination was needed for the entire flight of the Terrier and Tartar). This would allow the ASMS system to have more missiles in the air than its number of illuminators. Flight tests and laboratory tests, both conducted at the Applied Physics Laboratory (APL) and at General Dynamics Pomona, proved that Anderson was right.[4]

In September 1968 APL hosted a second SMS Technical Planning Group, which was convened on September 12 to update and extend the technical plan for the Surface Missile Systems that had been issued in 1963. The new report issued by Technical Planning Group addressed weaknesses in the 3Ts and emphasized the need for the ASMS. It stressed the need to automate targeting

* The Naval Ordnance Systems Command (NAVORD) was established on May 1, 1966, as a replacement for the Bureau of Weapons, which occurred when the Navy replaced the bureau system with the command system.

and detection via a Threat Responsive Weapon Control System and the importance of replacing analog fire control computers with digital computers.[5]

During the deliberations of Technical Planning Group II, it was perceived that with certain modifications the Standard Missile might be upgraded. In late 1969, just before the ASMS contract was awarded, the new missile that was originally included in the plans for ASMS was deleted and replaced with a modified version of the SM-1 with a strap-down inertial reference unit and a command communications link substituted in its place. This updated version of the Standard Missile was designated as the SM-2MR (later redesignated as RIM-66C).[6]

Prior to December 1969, establishing the operational and design criteria for the ASMS was the responsibility of the SMS Project Office (PM 3) in the Office of Naval Material. On December 18, 1969, the project was reassigned to NAVORD and redesignated PMO 403. Upon receiving approval to go ahead with the development of the ASMS defined by the Withington Committee, NAVORD initiated the acquisition process to begin procurement of the new system. By then, the Source Selection Evaluation Board had completed its final technical negotiations and had directed the three competing teams to submit their best and final offers. Based on these submissions, RCA's Missile and Surface Radar Division, located in Moorestown, New Jersey, was selected as the prime contractor. On December 23, 1969, it was awarded a contract to design and develop the ASMS to be called Aegis, after the shield of the Greek god Zeus, the ruler of the heavens. The name was suggested by Capt. L. J. Stecher, the Tartar System project officer who won NAVORD's internal contest for a short, catchy name for the project. As part of the contract, RCA would be responsible for designing the primary system radar, designated the SPY-1, an electronically steered, multifunction, planar array controlled by Navy's UYK-7 computers.[7]

The specifications for the SPY-1 radar and its control system—the key components of the Aegis Weapon System—emerged from contract studies conducted by APL, which had been assigned the task of validating the Withington Committee's recommendation after its findings were published in 1965. APL used the funding under the contract issued by the Bureau of Weapons (NAVORD's predecessor) to develop an experimental, digitally controlled high-powered phased-array radar called the Advanced Multifunction Array Radar (AMFAR). The new radar relied upon a phased-array antenna made up of several thousand individual phase shifters that were used to steer the radar beam electronically under computer control. This allowed the radar beam to be rapidly moved across its search area so that several hundred targets could be tracked simultaneously. As part of the AMFAR development effort, APL built a small phased-array antenna using a new type of ferrimagnetic garnet phase shifter that was engineered to overcome the temperature sensitivity of the conventional phase shifters previously used. The successful demonstration of the radar array used on AMFAR became the foundation for designing the phased-array antenna for the Aegis, SPY-1. This, coupled with the automatic detection and tracking performance of the AMFAR under wide variations of adverse conditions, formed the basis for the experimental development model (EDM-1) of the ASMS constructed by APL as part of its contractual obligations.[8]

CHAPTER 22

IMPLEMENTING THE AEGIS WEAPON SYSTEM

SHORTLY AFTER THE ESTABLISHMENT of the Aegis Project Office (PMO 403) on December 18, 1969, the head of the Naval Ordnance Systems Command, Adm. Mark W. Woods, called Capt. Wayne E. Meyer into his office. "We had a meeting last night," Woods informed Meyer, "and we selected you, Wayne, to be the manager of the new advance surface missile system." And that, according to Meyer's recollection, was how he became director of the Aegis project. He "didn't want anything to do with it," he said later in his oral history interview. "It would be nothing but grief and backbreaking. I didn't want it. I was in danger of losing my vocation when I saw how our Navy was inept or refused to look at situations that only it was capable of correcting."[1]

Meyer had been skeptical of the ASMS concept—which had now become Aegis—when he was briefed on it shortly after he had been made captain in the summer of 1966. Meyer, according to David Boslaugh, "was of the school that the best way to improve surface missile system performance was through evolutionary improvements to the existing Terrier, Tartar, and Talos system baselines, and he expressed unease with the ASMS approach because of his concern that it could easily be a repeat of the Typhon system that had tried to go too far beyond the bounds of achievable engineering practice." Now it would be his responsibility to see that Aegis worked. To make sure that this took place, Meyer would follow the doctrine of "Build a little, test a little, learn a lot" that he learned from Captain Spreen who had learned it from the developers of the Naval Tactical Data System (NTDS).[2]

At the time of his selection to head Aegis, Meyer was charged with overseeing a program that the authors of the Withington Study deemed to "be more difficult than any the Navy has faced, on any program, including Polaris." He had worked for a time on the Withington Study, and he used its findings and his own experience in the SMS project to prioritize the goals necessary to achieve the war-fighting capabilities set for Aegis. Meyer established five key priorities that would become the cornerstones of the engineering effort that led to the success of the Aegis Weapon System that Meyer designated as the Mark 7—a term inspired in part by the popular television program *Dragnet*, which always ended with a hammer stamping "Mark 7" productions. These priorities (in order of their importance) were reaction time, fire power, coverage, environmental immunity, and availability. Meyer believed that success could be attained only by paying constant attention to "people, parts, paper, and computer programs." As Robert Gray and Troy Kimmel, the authors of "The AEGIS Movement," point out, "One [of these priorities] could not be engineered in isolation from the others."[3]

Although he was selected as the Aegis project manager in December 1969, he did not report for duty until the following June, just in time to begin overseeing the $72 million dollars that had been budgeted for the 1971 fiscal year that was to begin on July 1, 1970. Meyer was immediately confronted by "a legion of naysayers that appeared almost before the project could get underway." Critics complained that RCA's phased array would cost too much, be too heavy, and consume too much power for the Navy's ships. Even the Defense Science Board—which was chartered to provide the Department of Defense with independent advice on science, technology, and acquisition—thought Aegis would be too complex, would cost too much, could not be operated by regular sailors, was not focused on the right threats, and could not be done on time.

The problems with the 3Ts and other "high technology" systems that had failed to live up to expectations had created a feeling among some members of the naval community that complexity and technology were bad. Many of these systems did not meet promised performance goals, and the contractors did not provide the kind of technical support, training, and spare parts necessary to keep the systems functioning. As former Secretary of the Navy John Lehman put it: "Industry went overboard in promising the moon and delivering green cheese."[4] "Some Navy people," reported the *Wall Street Journal*, "suspected that Aegis, which was named for a god in Greek mythology, was a recycled version of Typhon because it was named after the son of a Greek god." One of the programs greatest skeptics was Rear Adm. Henry E. Davies, deputy chief of naval material and development. "Programs at this stage develop a large constituency, and that gives them a tremendous inertia, making them really impossible to stop," he told the *Wall Street Journal*. "Belief in some of these new technologies becomes almost religious."[5]

Admiral Davies recommended against further development of the Aegis radar, claiming that potential benefits of the system did not outweigh the high costs of implementation. This put him at odds with the Navy's director of tactical electromagnetic programs, the director of Navy program planning, and the deputy chief of naval operatons for surface warfare, all of whom supported Aegis development and stressed the need to move the new system into the fleet. The technical specialists in the Aegis project office and their warfare sponsors in the Office of the Chief of Naval Operations (OpNav) considered Aegis too important to abandon. But the funds allocated to Aegis were consuming a great deal of the Navy's limited budget for engineering development.

This created a dilemma for Adm. Elmo R. Zumwalt Jr., the new CNO, who took office in the summer of 1970. When Zumwalt became CNO, many of the Navy's obsolete World War II destroyers were nearing the end of their useful lives just as the Soviet navy was deploying increasing numbers of advanced anti-ship missiles that were a serious threat to the Navy's carriers. The challenge facing Zumwalt was how to come up with a force structure that could protect the carrier battle groups from anti-ship missile attacks staged by Soviet aircraft and submarines. To counter anti-ship missile-carrying Soviet bombers, Zumwalt pushed hard for the F-14 fighter and its AIM-54 Phoenix missile. But the Soviet *Charlie*-class submarines that were being deployed with short-range SS-N-7 anti-ship missiles represented a more dangerous threat. These missiles could be fired outside the typical submarine detection range. Flying at Mach 0.9, an SS-N-7 would cover the thirty nautical-mile distance from its launch site to its target in about three minutes. To counter these missiles, the anti-aircraft batteries on the Navy's guided missile ships would have to have extremely fast reaction times. To protect the carriers, Zumwalt felt that it was necessary to balance the planned force of new frigates designed for anti-submarine warfare (ASW) with frigates designed for the anti-aircraft warfare (AAW) missions.[6]

As director of the CNO Systems Analysis Group—a post he held from 1966 to 1968—Admiral Zumwalt had established a study team to determine individual ship characteristics and force levels of the AAW and ASW platforms. The Major Fleet Escort Study issued by his office in 1967 proposed a family of three ship classes, designated as the DX, DXG, and DXGN, that would share many of the same weapon systems. The DX, the ASW version, was estimated to be the least costly to build at an estimated price of $24.9 million per ship based on a forty-ship build. The DX concept led to the design and construction of thirty-one DD 963 *Spruance*-class destroyers. The DXG, the AAW version, with a single Tartar D battery was estimated to be built at a cost of $42.6 million. Four examples of this type were built for the Shah of Iran but were never delivered. They were taken over by the Navy and converted into the *Kidd*-class guided missile destroyer. The DXGN was the AAW cruiser version that evolved into the nuclear-powered DLGN class. The Aegis system was initially intended for the first ship in this class, the USS *Virginia* (DLGN 38), which was projected to cost $256 million in 1969.[7]

After Admiral Zumwalt was sworn in as CNO on July 1, 1970, he set up a special project to "sidestep the bureaucratic" planning process that was hampering the Navy's efforts to modernize the fleet. It was named Project Sixty because it was tasked to submit its recommendations within sixty days. Historian Malcolm Muir termed Zumwalt's Project Sixty "the most effective strategic-planning exercise the Pentagon had seen since World War II. In it, the CNO redefined the Navy's four missions as deterrence, power projection, sea control, and presence." To solve the problem of what kind of escorts should be built to replace the Navy's aging surface escorts, the Project Sixty study group came up with a concept called the "high-low" mix. "High" was short for expensive high-performance ships having sophisticated weapon systems to protect the Navy's carriers; "low" was short for moderate-cost, moderate-performance ships that could be built in relatively large numbers for convoy protection. A high-low mix would allow the Navy to be in enough places at the same time to cover both sea control and power projection. The high cost of the Aegis system made Zumwalt's task of funding the large number of ships needed to fulfill the Project Sixty goals that much more difficult.[8]

When the *Virginia* was ordered on December 31, 1971, it was clear that the Aegis system would not be ready for installation on this ship as originally planned. Instead of the SPY-1 radar and its sophisticated control and command system, the Navy substituted the Tartar D missile fire control system and the SPS-48. It did include the newly developed Mk 26 guided missile launching system (GMLS), however. The modular design of the Mk 26 GMLS permitted the entire launcher with its magazine to be plugged into the ship. It also had an advanced fault-isolating system that enabled missile specialists to quickly identify problems. Because the Mk 26 could launch missiles in rapid succession under the worst sea conditions, its addition added greatly to the effectiveness of the *Virginia*'s AAW effectiveness. The Navy still expected to complete testing of the Aegis system in time for later ships of the DLGN 38 class, according to the testimony of Dr. Robert A. Forsch, the assistant secretary of the Navy for research and development; but this never came about once the program for the additional nuclear cruisers was scrapped.[9]

Although RCA had completed a cost-reduction study that had greatly simplified the design of the Aegis system while still retaining its essential features, Zumwalt wanted further cuts in the cost of the system. In December 1971, he asked if the Aegis system could be scaled down and procured at a lower cost. This request was passed to Meyer who dismissed it out of hand, advising Zumwalt that his office had already considered and rejected a scaled-down version of Aegis. Meyer's position was endorsed by Chief of Naval Material Command Adm. Isaac C. Kidd Jr. Admiral Kidd was aware of the importance of the Navy's guided missile ships, having activated the first all-guided missile unit in the Atlantic with five DLGs in 1962. He believed a scaled-down version was a waste of money.[10]

"At this stage," as Thomas C. Hone writes about Aegis and Rear Admiral Meyer, "Zumwalt considered canceling the whole project. He was angry because there was no AAW development plan to integrate the various ongoing AAW projects, and he correctly anticipated that Congress would resist funding sufficient numbers of an expensive, nuclear-powered Aegis ships." Rear Adm. Worth H. Bagley, Zumwalt's principal assistant, told Meyer that his boss was on the warpath.[11]

"He wants to kill something, and the Aegis is the nearest thing he's got to kill because that's where the money is."

"Well, what do you want me to do?" Meyer replied.

"Well," said Bagley, "we've got to show him something, my experience has been that you turn him around by letting him get exposed to something real, and he can't resist it."[12]

RCA had yet to start building the land test site for the Engineering Design Model-1 (EDM-1), but Meyer, who had selected APL to act as the program's technical adviser, knew about the experimental Advanced Multifunction Array Radar (AMFAR) stuff they had been working on before the contract. So, he arranged to schedule a demonstration of AMFAR equipment cobbled together to show the CNO. By luck, one of the technicians working at APL had been the chief fire controlman for the *Dewey* (DLG 14) when Zumwalt was in command. People were standing around and very pleased to see the CNO when he came to visit APL. And there, standing in the middle of the crowd was his chief fire controlman.

"Chief, what are you doing here?" Zumwalt asked.

"Well, Admiral, I work here now; I work for this."

"Chief, have you got anything to do with this Aegis?"

"Yes sir. That's where I'm working."

"Well," Zumwalt asked, "is it going to work?"

As Meyer recalled, "The chief pondered for a minute. He said, 'You know, Admiral, I think it will.'" And that was that. Zumwalt turned around, walked out, and was placated.[13]

After visiting APL, Zumwalt stopped attacking the Aegis program. It could not be faulted on the grounds of inefficiency (according to the Naval Audit Service), and the ongoing threat of the anti-ship missile could not be ignored. Canceling Aegis, Zumwalt concluded, would leave the Navy without any medium-range air defense and might threaten the future "high" capability surface escort program, which was then in the concept formulation and design stage. It took until November 1972 for Zumwalt to approve a production schedule for the Aegis radar and its control system, made with the proviso that Congress and the Navy had to agree on a platform to carry the new system. The failure to obtain a consensus within the Navy, and between the Navy and Congress, on whether Aegis ships should be nuclear or conventionally powered, however, caused further delays in the Aegis program. Hone asserts that the failure to define a suitable platform for Aegis "almost killed the system altogether."[14]

Meanwhile, RCA continued to make progress on EDM-1, the first of two EDMs that had been included in the original 1969 contract award of $252 million. This was a huge amount of money at the time, equivalent to more than $2 billion today. Included in the original contract was the requirement for RCA to design, build, and test an experimental EDM, identified as EDM-1, and to deliver a preproduction version identified as EDM-3. A third version, EDM-2, was to have supported the development of the ASMS missile at the White Sands Missile Range, but it was removed from the contract when the ASMS missile development was deleted from the program.[15]

RCA had begun to assemble the various components of the EDM-1 on its land-based test site in Moorestown, New Jersey, in March 1970. By the spring of 1973, it had completed the critical design review required by the contract, installed the SPY-1 radar at the land-based test site, and was tracking targets. RCA also submitted a detailed design for installing EDM-1 on the *Norton Sound* (AVM 1). The ship, which had previously served as a test bed for Terrier and other missiles, would be assigned the task of testing the EDM-1 in conjunction with the Standard Missile that was being developed concurrently by General Dynamics. In November 1973, during an operation called Fast Overdrive, Navy personnel operated and maintained the land-based system for two days. At the conclusion of the test, the EDM-1 was disassembled and moved by air to the Long Beach Naval Shipyard, where the *Norton Sound* was being converted to take on the engineering model of the Aegis system. The EDM-1, consisting of one face of the SPY-1 radar providing ninety-degree coverage, a single fire control illuminating radar, the new Mk 26 GMLS, the Aegis Radar Control Program installed on two UYK-7 computers, and the Aegis Radar Control Program installed on a single UYK-7 computer, was then installed on the *Norton Sound*. One hundred and twenty-six days after its arrival in Long Beach, the SPY-1 fitted to the *Norton Sound* began operating and tracking targets over the Pacific Ocean. On May 16, 1974, the Aegis EDM-1 automatically detected, tracked, and fired two SM-1 MR missiles, without warheads against a radio-controlled Firebee drone. The first missile successfully intercepted the target without destroying it, as planned. The second missile, fired seventeen minutes after the first, struck and destroyed the drone.[16]

Two months after the successful intercept demonstrated by the Aegis system installed on the *Norton Sound*, the Navy took an important step in overcoming the disruptive organization structure that had plagued its ability to integrate complex weapons systems into its ship designs. On July 1, 1974, Naval Ordnance Systems Command (NAVORD), which was responsible for developing Aegis and the Navy's other weapon systems, merged with Naval Ship Systems Command to become the Naval Sea Systems Command (NAVSEA). Meyer, who had been promoted to chief of NAVORD's Surface Missile System Division in 1972, was elevated to rear admiral and made head of the Aegis/SM-2 Weapons System Office (PMS-403) within NAVSEA with additional duty as director of NAVSEA's Surface Combat Systems, Group 3.* In addition to the Aegis ASMS Office, which Meyer had headed before the reorganization, Meyer was now in charge of the office responsible for the design of the Aegis cruiser and the destroyer-sized platform that was also being considered for the system. These organizational changes gave Meyer access to and control over ship design offices, direct contact with his sponsors in the Office of the Chief of Naval Operations, and potential contact with Congress. All these connections were needed if Meyer was to obtain the continued support and funding for Aegis. It also gave Meyer the ability to promote a new concept in the procurement of complex weapon systems called the Aegis Combat System. The new terminology was designed to describe all the elements that went into a ship, not just the Aegis system, but all its weapons and additional sensors as well.[17]

This idea came about after the engineers, striving to achieve the required minimum detection-to-launch time for the Aegis system, realized that the ship's command and control system would have to be contained within the Aegis system computers as opposed to resorting to a separate system such as the NTDS. If enough computing capacity were provided, they concluded, it might be feasible to control all the ship's fighting functions within the framework of a single system that was more cost-effective and easier to support. Meyer recognized the potential of this concept and used his new position to promote the concept of the Aegis Combat System that would expand Aegis from an air defense system into a multiwarfare, multimission system.[18]

Software was a major component of the Aegis system. It was assumed from the outset that so massive a program could not be created without software errors, so a degree of fault tolerance was built into the system. Meyer insisted that the system be designed specifically to deal with contingencies and future changes and additions. It was also essential that the system keep running despite damaged hardware or data errors. Thus, each element or module was designed to screen the data it received from other elements, and it had to be able to detect any errors and continue to operate despite some degradation of the system. The system was designed so that it could function without particular modules, in case of a casualty. Furthermore, it had to be able to isolate faulty equipment and reconfigure itself without stopping altogether.

* The assignment as Chief of Surface Missile System was a rear admiral's billet, which entitled Meyer to the prestige and authority of a rear admiral while not requiring the approval of Congress. He was officially promoted to rear admiral on July 1, 1975.

The above concepts were identical to those laid out by Cdr. Irvin L. McNally and Cdr. Edward C. Svendsen in the "Technical and Operational Requirements" document they prepared in 1955 for the proposed NTDS. To ensure the high reliability required in a naval weapons system, the two EDOs invented a design principle called "useful redundancy." They realized that if one critical element failed, it would bring the whole system down. To avoid this, each critical element was duplicated. "If one of the critical duplicated elements failed, the system would remain in operation but would gracefully degrade to a reduced capability using the remaining operable unit until the failed device was restored to operation." As the reader will recall, Meyer was responsible for integrating the NTDS system into the *Leahy's* modernization program and had agreed to use one of its computers for the daily system operating test (DSOT) program. Meyer undoubtedly picked up his ideas for Aegis from the NTDS documents.[19]

Meyer faced two major problems in his new post. The first was getting the Defense Department and the Navy's leadership to agree on what type of ship should be designed as an Aegis platform. The other was getting Congress to provide the additional funding needed to get Aegis to sea. On July 24, 1974, just a few weeks after Meyer's promotion, the Senate congressional budget conferees agreed to restore the $17 million in funding that the House of Representatives had eliminated from the $67 million Aegis budget requested by the Navy for fiscal year 1975. The House's action had been based on the belief that the Navy had not accomplished an acceptable level of system planning commensurate with the $400 million that had been expended on the program to that date. The conferees (from the House and Senate) decided to withhold future Aegis development funds unless the system demonstrated its ability to meet its prescribed performance objectives at sea while being maintained by "shipboard personnel only." Furthermore, such funding would be held unless the Navy and the Department of Defense agreed on a platform for Aegis and provided and integration plan specifying the interface of Aegis with the platforms.[20]

In January 1975, test firings against a variety of targets conducted by the *Norton Sound* were impressive enough to convince Secretary of the Navy J. William Middendorf II to release money that had been withheld pending the outcome of the sea trials that showed Aegis's tracking and fire control to be superior to that of any other AAW ship. But getting both Congress and the leadership in the Department of Defense to agree on an Aegis platform remained elusive, even though Meyer's project team had designed Aegis as a modular system so it could be fitted to variously sized ships (this was another idea that had been put forth by McNally and Svendsen). Adm. James L. Holloway III, who relieved Admiral Zumwalt as CNO on July 1, 1974, insisted that Aegis be installed on a 17,000-ton strike cruiser (CSGN) along with the Standard Missile, Tomahawk land attack missiles, Harpoon anti-ship missiles, a towed array, and LAMPS helicopters. He hoped to obtain eight of these ships that would be configured to carry out independent operations to take some of the strain off the hard-pressed carrier task forces. His predecessor, Zumwalt, also wanted Aegis, but he had preferred it to be mounted on a much cheaper hull than that of a nuclear cruiser, which was very expensive to build. Middendorf, facing severe budget cuts as funding for the military dried up after Vietnam, was also opposed to the CSGN, declaring that the expense of the Aegis system and a nuclear reactor were too great to bear.[21]

And then there was the problem with Congress. It was insisting that all major combatant vessels for the strike forces of the Navy be nuclear powered (unless the president advised Congress that the construction of nuclear-powered vessels was not in the national interest). When the Ford administration deleted a request for money for the lead nuclear-powered Aegis cruiser from its 1976 fiscal year budget, the Research and Development Subcommittee of the Senate Armed Forces Committee removed *all* the funding that had been allocated for the development of Aegis platforms. The elimination of funds to develop the ship types needed to field the Aegis Combat System posed a major problem for Meyer. Three project offices were involved in the Aegis program: PMS-403, the AAW system office; PMS-389, which was supposed to oversee procurement of a conventionally powered Aegis destroyer; and PMS-378, the nuclear-powered cruiser office. The congressional response to opposition to the nuclear cruiser promised to leave all three offices, which had grown in size in anticipation of the increased workload, without money.[22]

In May 1975, Chief of Naval Operations Holloway informed Secretary of Defense James R. Schlesinger that Congress would eliminate all Aegis funding if his office did not stand firmly behind some Aegis platform. At about the same time, the chair of the House Armed Services Committee wrote to President Gerald R. Ford in an effort to convince the president that all major surface combatants should be nuclear-powered, while simultaneously denouncing the influence of the "systems analysts" in the Office of the Secretary of Defense who were arguing against these costly ships. Adm. Hyman G. Rickover, the Navy's undeclared emperor of nuclear power, even suggested that the Navy propose the construction of a nuclear-powered destroyer without Aegis.[23]

"Aegis," to quote Tom Hone, "appeared doomed."[24]

Meyer, realizing that without funding his office would have to close down on August 1, 1975, wrote to the head of NAVSEA, Vice Adm. Robert C. Gooding, to request emergency funding to keep Aegis alive. While Meyer's request was under consideration, the Office of the Secretary of Defense proposed to senior members of the House Armed Services Committee that they provide separate authorizations for a gas turbine–powered Aegis ship and a nuclear-powered one. Aegis was eventually saved in September 1975, when a meeting of the House-Senate Conference approved funding for fiscal year 1976 to the amount of $45 million for conventionally powered and nuclear Aegis ships, and $66 million for the development of the radar/control system. Hone asserts that three individuals played a critical role in persuading the Conference Committee to restore funding for Aegis in the budget. One was President Ford, who promised to justify the need for a gas turbine ship in writing. Another was Vice Adm. James H. Doyle Jr., the deputy chief of naval operations for surface warfare. Doyle, a strong supporter of Aegis, convinced Holloway to support the idea of putting Aegis on a *Spruance*-class hull. Last, but not least, were the efforts of Meyer himself, who had "developed some pretty darn good relations" with congressional staff members, including Hyman Fine, a highly regarded member of the professional staff of the Senate Armed Services Committee.[25]

With funding assured for the next fiscal year, Meyer's office was able to begin construction of the Aegis Combat System Engineering Development Site (CSEDS), located in a former Air Force facility close to the RCA plant in Moorestown, New Jersey. It would house the engineering development model (EDM-3)—the production prototype—that would be used to validate the system's design before production was authorized. The original plan called for EDM-3 to be a development model for the weapon system alone. When the contract for this portion of the Aegis program was issued, RCA had been made the Combat System Engineering Agent for the Aegis ship. In addition to the Aegis system, the engineering agent was also responsible for the design and development of the ship's Command and Decision System. Thus, RCA became responsible for the functional and physical integration of all combat system elements installed on the ship in addition to the AAW elements of the Aegis system.[26]

When Meyer selected RCA as the Combat System Engineering Agent, RCA had already made a number of design changes and improvements in the phased-array radar, the software, and the support equipment. The changes in the SPY-1 radar were extensive enough to warrant its redesignation as the SPY-1A. The most important of these changes, from a cost standpoint, was the dramatic reduction in the price of the individual phase shifters, a critical component of the SPY-1 radar that enabled its single beam to be redirected almost simultaneously while being shaped as required. The phase shifters, an electronic device that changed the phase of the radio-frequency signals emitted by the SPY-1 antenna, were fed by traveling wave tubes that passed the desired wave forms created deep within the radar's electronics to the phase shifters via microwave tubing.

The phase shifters developed by APL used in the original SPY-1 phased-array radar antenna cost approximately $2,000 each to produce. A single phased-array antenna, such as the one used for the EDM-1, required a minimum of 4,096 phase shifters. Four such antennae would be required for each Aegis ship. The phase shifters alone would cost around $32 million per ship, which was 40 percent of the cost of the *Spruance*-class destroyer. RCA, recognizing the prohibitive cost of the phase shifters, embarked on a two-to- three-year effort with Raytheon to reduce their cost by redesigning the phase shifters. The redesigned phase shifters utilized a manganese-doped ferrite garnet secured from a small company located in Adamstown, Maryland, called Trans-Tech. Because of the company's unique capabilities, it was given responsibility for assembling the phase shifters themselves. Cost was the overriding design consideration faced by Trans-Tech. The finished product required a lot of manual dexterity and skill to produce, which initially resulted in a high rejection rate. Working closely with RCA and the Navy, Trans-Tech created a highly skilled assembly team using advanced manufacturing techniques that enabled them to provide the high yields necessary to achieve the cost reduction in the phaser shifters demanded by the Navy. The company was able to reduce the initial rejection rate of 40 percent to around 2 percent. According to those who were involved in the project, this reduced the unit cost of the phase shifters to approximately $200 each (according to Meyer, the final cost was about $40 each, but perhaps his memory was faulty).[27]

Other changes were made to the Air Warfare Weapon Control System, the radar signal processor, and the radar control program. The latter received extensive modifications to incorporate the lessons learned during the at-sea testing onboard the *Norton Sound* that increased the systems tactics to include electronic counter-countermeasures. The computer program for the Air Warfare Weapon Control System was also radically expanded to include a large number of missile engagement tactical situations not contained in the EDM-1, In addition to the air control function that was formally part of the NTDS, it was also expanded to include programming for the assignment of other ships' weapons in addition to coordinating the use of the Mk 26 missile launchers. This required the use of another four-bay UYK-7 computer as opposed to the single UYK-7 used for the EDM-1.

CHAPTER 23

GETTING AEGIS TO SEA

WHEN CONGRESS PROVIDED FUNDS in the 1976 budget to continue work on Aegis, it also provided funds to convert the nuclear-powered guided missile cruiser *Long Beach* (CGN 9) into an Aegis ship. Meyer and his design office now faced the dilemma of how to satisfy the Aegis design requirements to meet the needs of three different ship types: a strike cruiser, designated the CSGN; a destroyer type, the DDG 47; and a system to be retrofitted on the *Long Beach*.

Meyer's team solved this problem using a design and engineering approach called Superset. The name of the individual who came up with this idea has been lost to history, but those who are familiar with this aspect of the Aegis story consider it "a brilliant scheme." The philosophy underlying the Superset scheme is based on the idea of first integrating the largest set of system elements (weapons, sensors, control systems, etc.) under consideration, and then down-sizing the Superset capabilities to fit the needs of a smaller platform. The largest configuration of the Aegis Combat System was prepared for the proposed CSGN strike cruiser. Those parts of the Superset not needed for the DDG 47 destroyer or *Long Beach* conversion were dropped from the Aegis Combat System configuration on each of those ships.[1]

As Tom Hone points out, "Friction between the Whitehouse and the Navy, and conflicts between then President Jimmy Carter and Congress, continued to threaten the success of the Aegis program." The Carter administration continued to make changes in the Aegis ship types. It withdrew funds for retrofitting the *Long Beach* from the fiscal year 1978 budget and canceled the CSGN. The latter was replaced with a follow-on to the *Virginia*-class guided missile cruisers designated CGN 42. The Carter administration originally planned to build sixteen DDG 47s and eight CGN 42s, but funding for the CGN 42s was eliminated from the following year's budget, leaving the DDG 47 as the only remaining Aegis platform.[2]

The DDG 47 was intended to be a repeat of the highly successful design used to construct the *Spruance*-class destroyers modified to incorporate the Aegis Weapons System. The *Spruance*-class were the first U.S. Navy warships equipped with a gas turbine propulsion system. Although they were as large as contemporary cruisers, they lacked the armament necessary to be designated as cruisers. In 1974, *Spruance*'s builder, the Ingalls Shipbuilding Company, presented Meyer with a proposal to install Aegis on a *Spruance* hull form. After being rejected by a Navy engineering review, Meyer commissioned an independent feasibility study of installing Aegis on a *Spruance* hull through RCA, which then issued a contract for the study to the naval architecture firm of John J. McMullen Associates. The McMullen study, reviewed by the Navy's ship engineering community, showed that it was feasible to install Aegis in such a hull, provided significant design changes were made to the hull. This study, according to those familiar with it, "moved the DDG 47 option to center stage as the earliest practical path for deployment of AEGIS."[3]

As Meyer's team was engaged in formulating the various design sets required to get Aegis to sea, some of those holding positions of authority in the upper ranks of the Navy's leadership were beginning to recognize the need to reorganize the administrative structure within the Navy used to procure ships and weapon systems. For most of its history, the Navy acquired its ships and weapons independently of each other. Under the bureau system, first established in 1842, ship design and procurement fell under the purview of the Bureau of Construction and Repair;* weapons and ammunition under the purview of the Bureau of Ordnance. When the system commands replaced the bureaus in 1966, these functions were shifted to the Naval Ship Systems Command and the Naval Ordnance Command. Although these commands were merged into the Naval Sea Systems Command in 1974, decisions concerning the ships' characteristics and their offensive and defensive systems continued to reside within the Surface Warfare Division within OpNav. As a result, problems with the coordination and integration of weapons continued to plague the Navy.

Meyer was highly critical of the design of the recent nuclear cruisers. In his opinion, the sensors and weapon systems on these ships were poorly integrated,

---

* The Bureau of Construction and Repair and the Bureau of Engineering were combined to form the Bureau of Ships in 1940.

they lacked the capability to manage a battle group's anti-air and anti-submarine information and would have difficulty managing a battle group's weapons in major engagements. He was acutely aware of the problems of trying to integrate complex weapon systems within the current bureaucratic structure, and he worked hard trying to convince his superiors that this was not the best way to design and build the Navy's warships.[4]

Meyer found an ally for this effort in Vice Adm. James H. Doyle Jr., the deputy chief of naval operations for surface warfare. Doyle, a nuclear-trained officer with a law degree, took over the Surface Warfare Office (OP 03) within OpNav on September 30, 1975. His responsibility for surface ships and their weapon systems brought him into immediate contact with Meyer, who provided Doyle with frequent updates on Aegis and the SM-2 program. The two officers met frequently on a day-to-day basis to discuss the operational and technical planning for future Aegis ships, how to staff Meyer's office, and how to integrate the leadership of OpNav and Meyer's PMS-403 within the Navy's Material Command.[5]

Even though Doyle had not met Meyer personally before becoming the deputy chief of naval operations for surface warfare, he held Meyer in high regard due to an incident that occurred in April 1969 while Doyle was in command of the *Bainbridge* (CGN 25). The *Bainbridge* had been in Subic Bay in the Philippines for maintenance when Captain Meyer, then technical director of Naval Ship Missile Systems Engineering Station (NSMSES) at Point Hueneme, California, came aboard to inspect the ship's Terrier system. Although Doyle was away from the ship during Meyer's inspection, his weapons officer reported that "Meyer was upset with what he found and told the officers in no uncertain terms what must be done to correct the situation." Doyle's training in Admiral Hyman G. Rickover's highly demanding nuclear training program made him a stickler for seeing that written procedures were carried out. Thus, his reaction to Meyer's action was extremely positive. Doyle considered the event the beginning of "a strong personal bond" between them.[6]

A major problem facing Doyle upon taking over OP 03 was the ongoing fiasco surrounding the Navy's recent shipbuilding activities that was based, for the most part, on the use of Total Package Procurement Contracts. Such contracts gave private industry the luxury of having only to satisfy broad performance goals and requirements. According to Doyle, "This approach, plus fixed cost contracts, arbitrary schedules, unrealistic share lines, and speculative cost estimates, resulted in billions [of dollars] in claims" against the Navy. Meyer was against fixed contracts, too, which seems to have pitted him against Secretary of the Navy Lehman.[7]

One of Doyle's first tasks as head of OP 03 was to explain to Secretary of Defense James Schlesinger why there were such huge claims against the Navy. The use of Total Package Procurement Contracts, Doyle explained, put industry in the driver's seat. Mission degrading discrepancies during acceptance trials had forced the Navy to issue acceptance waivers that led to additional costs and second construction periods to correct the deficiencies. The current procurement system, according to Doyle's analysis, had numerous problems that needed to be fixed. Among the issues that needed to be addressed were the following:

- Leadership and management, financing, and budget control were dispersed among several institutions with limited coordination.
- Hull construction dominated weapons and armament development.
- Various equipment and elements were engineered, built, and supported as separate commodities.
- Combat systems, to the extent they could be labeled such, were wired together by the shipbuilders without qualified workers or adequate documentation.
- New, emerging computer programs were incomplete, with limited testing and very limited in-service support.
- Crews had no active role in the builder's and acceptance trials.

"In short, there was no one person in charge in the Navy with the requisite authority, expertise, and resources to harness government and industry as a team engineer and execute the shipbuilding program."[8]

Doyle, having made "getting Aegis to sea" his number one priority, was determined to resolve the divided accountability problem that existed between Meyer's Aegis Weapon System Office and the surface ship managers (one for cruisers and another for destroyers) in the Naval Sea Systems Command (NAVSEA). With Meyer's help, Doyle convinced the commander of Naval Material Command, Vice Adm. Frederick H. Michaelis, and the commander of Naval Sea Systems Command, Vice Adm. Clarence R. Bryan Jr., to establish a single project office responsible for engineering, design, development, production, testing, introduction, and in-service support of all Aegis Combat

Systems and Aegis ships, with the requisite control of the allocated budgets. It was named the Aegis Shipbuilding Project (PMS-400). When the PMS-400 office was stabled in June 1977, Admiral Bryan appointed Rear Admiral Meyer as project manager.[9]

As head of the Aegis Shipbuilding Project, Meyer now had control over the development and production of the Aegis Weapon System, all versions of the Standard Missile, the Aegis Combat System, and the building of all Aegis ships. To fulfill these responsibilities, he had to acquire and organize a staff, prepare designs for contractors, develop a working relationship with his sponsor in OpNav, make sure the Aegis ships met fleet needs, and keep Aegis afloat in Congress. He began by drafting the new organization's charter, which was prepared with Admiral Bryan's permission. The charter made Meyer responsible directly to the chief of NAVSEA, authorized Meyer to act on his own initiative in any matter affecting the project, named Meyer the delegated authority of the chief of naval material, and centralized control over Aegis ship procurement and Aegis system development in PMS-400. This made Meyer fully accountable for Aegis ship acquisition, gave Meyer responsibility for preparing and signing the fitness reports and performance ratings of all military and civilian personnel assigned to PMS-400, made Meyer responsible for total ship system engineering integration, and gave PMS-400 the duty of integrating all the logistics requirements for Aegis ships.[10]

While Meyer worked to organize and integrate the staffs from PMS-403, PMS-378, and PMS-389 into the PMS-400 office, the Navy continued to clash with the Secretary of the Navy and President Carter over the future size and composition of the Navy. A month before Meyer was appointed director of PMS-400, Carter announced that his administration would authorize 160 new ships for the Navy over the next five years. A year later, President Carter, at the urging of Secretary of Defense Harold Brown, cut this figure in half, casting doubt on funding for the lead DDG 47. Adm. Thomas Hayward, the new CNO, strongly opposed Secretary Brown's views on the size of the Navy. The admiral appeared before the armed services committees of both houses of Congress to argue that the Navy needed the advanced AAW capability that only Aegis ships could provide. Without Aegis, he testified, existing carrier battle groups would be at great risk in the 1980s. His case was helped by a public rift between Brown and Secretary of the Navy Graham Claytor, who complained bitterly about the Carter administration's spending for the Navy.[11]

Meyer's ability to convince members of Congress that Aegis would work was responsible for assuring that funding for the first Aegis destroyer remained in the 1978 fiscal year budget. One former member of his PSM-400 staff referred to Meyer as a "master communicator." Contributors to the Naval War College's *Politics of Naval Innovation* noted that he "targeted his key audiences and dealt with each accordingly." To deal effectively with Congress, he studied the key members of the appropriation committee, intent on learning their views about the Navy's ships and weapons. He noted the time of the budget process, and when necessary, bypassed the Navy's Legislative Affairs Office.[12]

To win support within the Navy, Meyer brought representatives from many Navy shore activities into the Aegis program by "double hatting" them, that is, giving them positions of responsibility within PMS-400 in addition to their regular Navy jobs. This gambit created teams of Aegis advocates in the Navy's shore-based organization and in OpNav. Meyer also took a politically aggressive approach when it came to problems emanating from the Office of the Secretary of Defense, and he viewed the Defense Systems Acquisition Review Council (DSARC) process as an opportunity to discipline the major contractors.[13]

In 1975, the DSARC directed the Navy to incorporate the Aegis Weapon System into the DD 963 hull making only those changes to the hull necessary to install Aegis's mechanical and electrical systems and to correct any known deficiencies in the DD 963 design. To accommodate Aegis, it was necessary to raise the bulkhead deck from the main deck and to redesign the superstructure to accept the SPY-1 radar. Other superficial hull changes included relofting the sonar dome, the addition of a bulwark on the forecastle, and widening the hangar to accept two LAMPS Mk II helicopters. These and other changes to the mechanical and electrical systems yielded a preliminary design estimate for a ship having a displacement of 9,043 tons. This figure was reduced to 8,910 tons during the concept design phase.[14]

As project manager, it was Meyer's responsibility to provide the DSARC via the Decision Coordination Paper with the information it needed to make its recommendations to the Secretary of Defense. The Decision Coordination Paper was needed to start the acquisition approval process, and it was issued at the beginning of a project or at each new phase of the acquisition. Meyer's office had to provide the information needed to prepare the Aegis/DDG 47 Decision Coordination Paper.[15]

Meyer finally received approval to proceed with the contracts for the Aegis Weapon System and the DDG 47 destroyer after the third Aegis DSARC gave its blessing to the program in January 1978. With money and congressional support in hand, Meyer took charge of the critical task of building the Aegis fleet. In April, RCA signed a contract with the Navy in the amount of $226 million for the first Aegis Weapon System Mark 7 Mod 3 that was destined for installation in the first Aegis ship. Ingalls Shipbuilding, a division of Litton Industries in Pascagoula, Mississippi, won the competition to build the DDG 47 and in September 1978 was awarded a contract for its detailed design and construction valued at $287.8 million. During the award ceremony, which was conducted in Washington, D.C., Rear Admiral Meyer in his address to the gathered guests described the new ship as "the most capable, heavily armed and survivable destroyer† the Navy has ever built."[16]

At the time of the contract award, on September 22, 1978, both Ingalls Shipbuilding and NAVSEA agreed that the design for the DDG 47 was "a very tight ship" with regard to weight and center of gravity requirements. When Ingalls submitted its baseline weight estimate in February 1979, it was 360 tons greater than the Navy's contract design estimate of 8,910 tons. Both sides agreed that every effort should be made to reduce the weight and the center of gravity. This was needed to improve the ship's stability. Additional weight would increase displacement causing the ship's hull to ride lower in the water. Raising the center of gravity, a problem created by the increased height and weight of the Aegis super structure (dictated by the SPY-1 antennae), affected the ship's metacentric height and its stability. To avoid these problems, the Navy initiated a weight reduction program in June 1978. Despite these efforts, the ship continued to grow in size.[17]

The keel for the first Aegis ship was laid down at Ingalls Shipbuilding's shipyard in Pascagoula, Mississippi, on January 27, 1980. Twenty days earlier, her designation had been changed from that of a destroyer bearing hull number DDG 47 to that of a cruiser with hull number CG 47. Although the ship had been designed around the hull of a *Spruance*-class destroyer, the ship's greater displacement of 1,225 tons, along with greater fire power and facilities for a unit commander, warranted the upgraded classification as a cruiser. The ship was built in sections called modules that were constructed from subassemblies

† DDG 47 (USS *Ticonderoga*) was redesignated as a cruiser, CG 47.

outfitted with piping, ventilation ducting, and other hardware needed to fulfill the shipbuilder's requirements. The modules were then moved together to form the hull of the ship, and the deckhouse sections were then lifted aboard. For launch, the ship was moved several hundred yards over land via a wheel-on-rail transfer system to the floating dry dock, which was used to launch the ship. It was floated off the dry dock on April 25, 1981, and moved to an outfitting berth. The ship was christened USS *Ticonderoga* by First Lady Nancy Reagan on May 16, 1981.[18]

CHAPTER 24

THE TROUBLESOME *TICONDEROGA*

TICONDEROGA'S (CG 47) DISPLACEMENT as delivered was 9,589 tons. The increase in her design displacement over that of the *Spruance* created a public relations nightmare when investigators on the staff of the U.S. House of Representatives Appropriations Committee released a report criticizing the stability and speed of the ship without understanding the measures that the designers (who had discovered the problem long before the congressional audit team) had taken into account for the added displacement. Although the *Ticonderoga*'s draft was six inches greater than the *Spruance*'s, instability was never an issue based on the addition of 200 tons of ballast and other design changes instituted by her builder. But when the report was leaked to the press, it created a firestorm that spread across newspapers and magazines throughout the country. One journalist went so far as to claim that the ship would capsize in heavy seas and was "so burdened with technological lard that it [could] not keep up with the other warships of the Navy's carrier battle groups." Eventually the storm subsided thanks to the efforts of Meyer's team, and work continued on the outfitting of the *Ticonderoga*.[1]

Criticism of Aegis, which was one of the nation's most expensive weapons, continued to plague the program. After *Ticonderoga* was commissioned on January 22, 1983, the General Accounting Office (GAO) issued a report, which was picked up by newspaper columnist Jack Anderson, saying that Aegis had never been tested against low-flying cruise missiles that were considered one of its more lethal threats. In subsequent hearings, Vice Adm. R. L. Waters, the deputy chief of naval operations for surface warfare, testified that tests against low-flying targets had been conducted with the Aegis system installed on the test ship *Norton Sound* (AVM 1) "against targets representative of every known threat or expected threat expected to be deployed by the Soviets in this decade."[2]

Senator Gary Hart on the Senate Armed Forces Committee was highly critical of Aegis and stated that he would block funds until a new series of tests "designed by CG 47 Aegis critics, not the Navy," had been conducted. Meyer, as reported in the *Wall Street Journal*, was not going to accept that. "All that success-fail criteria ain't gonna mean a damn thing except to the technical people," he exclaimed acerbically to the testing unit that was preparing to give the *Ticonderoga* a series of simulated attacks off Puerto Rico in April. Nevertheless, he revised some of the test criteria (according to the *Wall Street Journal*) to make them less demanding.[3]

The results of the tests, according to the *Wall Street Journal*, "bordered on the fantastic," based on the Navy's claim that the *Ticonderoga* had been attacked thirteen times by drones that popped up over the horizon and were all shot down. The official report presented to Congress a year later was no less glorifying in the test results of the Aegis system conducted by the Operational Test and Evaluation Force:*

> Operational testing is a critical element of our decision-making and Congress had mandated important organizational changes at the OSD [Office of the Secretary of Defense] level to ensure that adequate attention and effort are devoted to planning, analysis and objective appraisal of the results of operational tests of new systems before rate production is approved. Our independent testing agent, COMOPTEVFOR, [Commander Operational Test and Evaluation Force] continues to provide objective professional appraisals of the strengths and weaknesses of new naval systems. The AEGIS system on the TICONDEROGA class cruisers is an excellent example of a system that had been subjected to this rigorous and impartial testing. The Navy conducted both developmental and operational testing of the USS TICONDEROGA (CG 47) and the AEGIS weapon system in April 1984. The purpose of this testing was to verify the correction of deficiencies identified in previous operational testing. The testing was more comprehensive and demanding than any undertaken to date. Multiple and simultaneous attacks by unmanned targets simulating some of the world's most sophisticated anti-ship missiles were conducted. Most attacks were in a wartime jamming and chaff environment. The USS TICONDEROGA countered these attacks with live warhead missiles and a wartime firing doctrine. The AEGIS weapons

* The Operational Test and Evaluation Force is an independent agency within the Navy charged with providing an independent and objective evaluation of ships, aircraft, and weapons systems.

> system performed magnificently, destroying ten of eleven targets during operational tests and five of five targets during development tests which immediately preceded the operational phase. These tests proved that our confidence in the AEGIS system was completely justified.[4]

Ticonderoga was an extraordinarily advanced ship at the time of her commissioning; she went to sea on time and within budget. She retained the original destroyer maneuverability of the DD 963 hull, and at 25 knots could turn 90 degrees in two ship lengths. A steel bulwark, switched in later ships to aluminum to reduce top weight, was added to the bow to keep the focsle dry in moderate seas, as the cruiser pitched hard in heavy weather. Her four SPY-1 radar arrays were mounted in two enlarged deckhouses, which became the hallmark of her outboard profile.[5]

In addition to the Aegis Mark 7 Weapons System, the *Ticonderoga*, at the time of her commissioning, was equipped with two Mark 45 5-inch/54-caliber lightweight guns, two Mark 15 Phalanx close in weapons systems, two Mark 32 12.75-inch triple torpedo tubes, two Mark 8441 guided missile launchers each containing four RGH-84 Harpoon missiles, two SH-2F LAMPS helicopters, one SPS-49 air search radar, one AQO-9 gun fire control radar, an SQS-53A sonar in the bow, and an SLQ-32(V)3 Electronic Warfare Suite. The Aegis system was composed of eight major elements: the SPY-1A phased array radar, the Command and Decision System Mark 1, the Weapons Control System Mark 1, the Fire Control System Mark 99, the Guided Missile Launching System Mark 26, the Aegis Display System Mark 1, the SM-2 MR Block I missiles, and the Operational Readiness Test System Mark 1.

The heart of the Aegis Weapon System installed on the *Ticonderoga* was its high-power, multifunction radar consisting of four, computer controlled, phased-array antennae strategically located to provide full hemispheric coverage around the ship. The SPY-1 radar could scan the sky in all directions at once to find and track aircraft and missiles while continuously watching for new targets. Each of these 12 × 12–foot flat antennae contained 4,000 radiating elements that could direct pencil beams of high power in one direction, jump to another direction, or split energy in several directions. The combination digital control, coupled with the phased-array elements, enabled the Aegis radar to search the volume around the ship at varying rates, depending on the specific tactical objectives selected by the operator. For example, searching the horizon was usually faster than a searching above the horizon to detect low-flying cruise missiles. Once the system detected an object, it was instantaneously placed in the precision automatic monopulse tracking mode, and the results automatically fed to the ship's Mark 1 Command and Decision System.[6]

The primary function of the Command and Decision System installed on the *Ticonderoga* was to perform threat evaluation and weapons assignment. Taking information from the SPY-1 radar and the other sensors onboard, it determined whether a threat existed, how serious it was, and what to do about it. If the threat needed to be destroyed, it formulated an engagement order that was passed to the Mark 1 Weapons Control System. This information was passed to the operators, or it could be used in the system's automatic mode to act independently without human intervention. To prevent the interception of incoming and outgoing combat air patrols, special zones were semiautomatically established according to the carrier's flight paths. In practice, only semiautomatic modes were employed. In most cases, the system was used to indicate increased alert zones from which an attack would most likely be launched against the battle group.

Using track data from the SPY-1, the Weapons Control System determined which target was engageable, selected the appropriate launcher and missile, pointed the launcher in the launch direction, and initiated the launch sequence for an SM-2 missile. After launching the missile, the system continued to track the missile and the target via the SPY-1 radar, generating steering commands transmitted to the missile via the SPY-1 during its midcourse flight. As the interception point, the Weapons Control System sent seeker-pointing commands to obtain target acquisition via the illuminator signals reflected from the target. Data from the Command and Decision System was also passed to the Mark 1 Aegis Display System, which on *Ticonderoga* consisted of four large-screen displays that formed the nucleus of the ship's combat information center. The display system geographically portrayed the tactical environment surrounding the ship. The system summarized status information on the resources available and presented it in a format that aided the decision making by the commanding officer and his tactical action officer. The display system was not part of the original Aegis Weapon System but was added to the system when OpNav realized that the vastly improved air picture provided by the SPY-1 would make the Aegis ship the natural candidate for the role of Battle Force Anti-Air Warfare Coordinator. The Mark 99 Fire Control System was another essential component of the Aegis Weapon System. It was composed of four dish-shaped transmitters, each mounted on a two-axis director, and the digital software to control them. The fire control system illuminated the target with X-band radar provided by a powerful transmitter located below decks to support missile homing.[7]

Another module included in *Ticonderoga*'s Aegis Weapon System was the Mark 1 Operation Readiness and Test System (ORTS). This system was designed to continuously monitor the health of the Aegis system by continuously monitoring 2,000 static test points built into the system and many of its operability tests. As noted by Joseph Threston, an RCA employee who was in charge of Aegis for twelve years, "Once a fault is detected, the ORTS will isolate the fault to the lowest replaceable unit. It then provides, via electronic displays, a list of the repair parts required and the instructions necessary for the sailor to make the repair."[8]

As Vice Admiral Doyle writing in the 2009 issue of the *Naval Engineers Journal* has noted,

> The total system of people, parts, paper, and computer programs was defined. Critical functions were logically, physically and managerially centralized. Teamwork was created among designers, engineers, trainers, logisticians, and shipbuilders. Laboratories and field stations were aligned with industry. Close communications were established with the shipbuilders and the combat system agent. Aegis Area Commanders were established at key shipyards, industrial plants, land-based test sites, and major combat systems engineering centers. Plans and budgets for in-service engineering teams were developed for lifetime support. The introduction of combat system baseline engineering, together with computer program development, system production and test, crew training, propulsion testing, hull construction, and ship integration and test, were all put in place.[9]

How Meyer achieved this success is laid out in Hayes Smith's edited volume *The Politics of Naval Innovation*, published by the Naval War College in 1994. The seven key points presented by the authors of the chapter on Aegis are as follows:

1. Making PMS-400 field representative de facto deputy program managers, so that contractors dealt regularly with an office possessing real authority.
2. Travel, with frequent on-site inspections and reviews. According to one witness, Meyer could be "ferocious" in these reviews, particularly of RCA

and Ingalls. But his goal was to make adhering to production schedules a matter of pride. As one former staffer in PMS-400 said, "Meyer loved to kick the tires." That meant lots of visits, even to subcontractors. RCA, for example, used PMS-400 to discipline Raytheon, one of its major subcontractors. And Meyer traveled regularly to smaller contractors, handling out efficiency awards and exhorting quality work.

3. Testing in parallel with production. PMS-400 "tested the hell out of the system," according to a former Operations Division director, because Meyer did not want any surprises. His goal, after all, was to produce a revolution in naval weaponry, and he was determined to turn his vision of warfare into a working reality.
4. Not allowing PMS-400 to become captive to routine. *All* the former staffers of PMS-400 interviewed [by the authors] said Rear Admiral Meyer was a very demanding manager. Yet all respected him. They admired his fierce concern for excellence. As he admitted himself, "I harped on that and harped on that from day one." They also admired his willingness to listen. One noted that Meyer was often not sure how to translate his "visions" into reality, so that senior contractor personnel wasted lots of time on ideas that did not pan out. But work was never dull. Meyer tapped key PMS-400 junior staff to answer congressional questions and write speeches, and senior staff to hand out "Aegis Excellent Awards." About every six weeks, the admiral called a halt to travel, stuffed all of PMS-400 into a conference room, and reviewed the project's status. He also gave out awards and "fired up the crowd." Then it was back to travel and meetings.
5. Getting practical control of much of his contractors' organizations. Meyer reached around RCA and Litton [the owners of Ingalls] management to communicate with the people doing the work. Meyer also used the Applied Physics Laboratory (APL) and a number of independent consultants to review both the technical and managerial practices employed by his major contractors. His goal was to create a community of Aegis supporters and experts. As one of Myer's former deputies put it, "Meyer build a national organization through his prime contractors."
6. Keeping fleet organizations informed with briefings, newsletters, films, and demonstrations. The Combat Systems Engineering Development Site (CSEDS) was used to show high ranking Navy officers and

influential members of Congress what Aegis could do, but it was also turned into a training station for AAW software development. To Meyer, Aegis was not a static system, and the heart of its "evolution" was its software. CSEDS both modified the software and showed it off. PMS-400 also planned programs to maintain and modernize Aegis ships.

7. Justifying Aegis to keep potential opponents quiet. All responsible personnel in PMS-400 were tasked with defending Aegis against criticism. In the process, they often-anticipated real problems and potential criticisms; the justification process was itself a planning tool.[10]

Despite the huge success he had achieved, Meyer was removed from his job as Aegis project manager in August 1983 and assigned to duty as the deputy commander for the Combat System Directorate. It appears from statements in Meyer's *Reminiscences* that he and Secretary of the Navy John F. Lehman Jr. did not get along very well. "I mean we'd come to a point and I had no use for the man." Meyer claimed that Lehman sent him to the directorate "because it was a dead end, it was someplace out in the wilderness, as a way to get rid of me without him being accused of mistreating me."[11]

Meyer's recollection is troubling based on the magnitude and importance of the Combat System Directorate, which is responsible for managing research and development, acquisition, maintenance, engineering, logistic support, and material management for all the projects and systems assigned to the Combat System Directorate (table 24.1). Also troubling is Meyer's failure to be promoted to vice admiral. His name was on the list of those selected to receive their third star, but Meyer was denied promotion when none of the three EDOs on the selection board voted for his promotion. There was often an acerbic tone in his remarks denigrating certain senior officers that he had encountered, various structures within the Navy, and its procurement procedures in general. It also shows that he was not afraid of criticizing anybody or any part of the Navy not up to his standards. It is likely that Meyer made a lot of enemies on his way up.[12] In any case, toward the end of his second year as deputy commander of the Combat Systems Directorate, he had had enough and put in his papers for retirement. Neither the Secretary of the Navy nor the current CNO were available to speak when Rear Adm. Wayne Meyer retired from active duty on December 1, 1985, at the age of fifty-nine after forty-two years of service.[13]

TABLE 24.1

COMBAT SYSTEMS DIRECTORATE, NAVSEA 06

PROGRAMS, OFFICES, AND PROJECTS, 1983
NATO Seasparrow Program Office
Combat Systems Engineering Research and Technology Office
Class Combat Systems Engineering Subgroup
Combat Systems Design and Test Subgroup
Combat Direction Systems Subgroup
Combat Support System/Equipment Subgroup
Engineering and Logistics Support Office
Surface Warfare Systems Research and Technology Office
Surface to Surface Missile Weapons Systems Subgroup
Surface Gun Weapons Systems Subgroup
Surface Missile Weapons Systems Subgroup
Advanced Systems and Integration Office
Engineering Support Office
Submarine Systems Subgroup
Surface Systems Subgroup
Undersea Weapons and Test Subgroup
Ammunition Management Division
Aegis Shipbuilding Project
Torpedo MK 48 Project
Directed Energy Weapons Project
Advanced Lightweight Torpedo Project
Mine Warfare Systems Project
Naval Shipboard Tactical Embedded Computer Resource Project

Source: U.S. Navy, Naval Sea Systems Command, *Contracting Opportunities,* 2nd ed. (Naval Sea Systems Command Liaison Branch, 1983), III.16–III.41.

CHAPTER 25

REICH AND MEYER

EXCEPTIONAL SERVICE IN THE U.S. NAVY

ON OCTOBER 18, 2008, THE 58TH *ARLEIGH BURKE*–class Aegis destroyer was christened the *Wayne E. Meyer* (DDG 108) in a ceremony at Bath Iron Works in Bath, Maine. Mrs. Anna Mae Meyer served as sponsor of the ship, named for her husband, who led the development of Aegis, the first fully integrated combat system built to defend against air, surface, and subsurface threats. The *Wayne E. Meyer* entered the U.S. Navy a year later, on October 10, 2009, when it was commissioned during a ceremony at Penn's Landing, Philadelphia, Pennsylvania.

The naming of U.S. Navy destroyers can be traced to the Spanish-American War, when then–Assistant Secretary of the Navy Theodore Roosevelt made the decision to name the first sixteen torpedo-boat destroyers after officers and enlisted men of the U.S. Navy and Marine Corps. This naming convention carried over to the destroyers, a ship type that evolved from the earlier torpedo-boat destroyers. This broad convention allowed the Secretary of the Navy to name ships in honor of distinguished leaders and heroes from each of the services. By 1942, the convention had been modified several times to include "deceased American Naval, Marine Corps and Coast Guard Officers and enlisted personnel who have rendered distinguished service to their country above and beyond the call of duty; former Secretaries and Assistant Secretaries of the Navy; members of Congress who have been closely identified with Naval affairs; and inventors."[1]

The practice of naming warships only for deceased Americans or heroes continued after World War II until the Riera Panel of 1969, which decreed that "vessels will not be named to honor living persons," and endorsed an earlier decision to exclude members of Congress from the destroyer naming convention. This decision had unforeseen consequences for orthodox traditionalists. Simply put, there was simply no way that future Secretaries of the Navy would fail to recognize the World War II and Cold War legislators who helped create the most powerful naval force on earth. The Riera Panel's decision compelled the secretaries to honor the Cold War generation of leaders by using exceptions to established naming conventions across different ship types. It also signaled a return to the Continental Navy tradition of occasionally honoring famous living persons with a ship name as exemplified by the decision to name DDG 51, the first Aegis destroyer, for the living World War II hero Arleigh A. Burke.[2]

Rear Adm. Wayne Meyer had been the founding project manager for Aegis shipbuilding and was affectionately known as the "Father of Aegis" for his untiring efforts in securing the resources necessary for the successful development of Aegis, which some claim "transformed the nature of naval warfare." In honor of Meyer's work on Aegis, the Navy, in 2006, announced that it would name its newest *Arleigh Burke*–class guided missile destroyer after him. Two years later, in October 2008, Meyer proudly looked on as his second wife* christened the destroyer bearing his name with a bottle of champagne. He died of congestive heart failure at the age of eighty-three on September 1, 2009, one month before the ship was commissioned.[3]

Although Vice Adm. Eli T. Reich played an important role in laying the groundwork for Aegis and achieved a higher rank than Meyer, his efforts were less publicized and remain largely unknown. Nevertheless, his contributions to the development of the organizational changes within the Navy that enabled the successful implementation of Aegis were recognized when the Navy named the facility for testing the Navy's combat systems and electronic equipment at the Naval Ship Weapon Systems Engineering Station at Port Hueneme, California, Reich Hall.

* His first wife, Margaret, predeceased him.

APPENDIX

CHRONOLOGICAL COMPARISON OF CAREERS

	ELI T. REICH		WAYNE E. MEYER	
Date	**Rank**	**Duty**	**Rank**	**Duty**
Jun 35	Ensign	Graduates USNA		
Jun 35–Feb 36	Ensign	*Pensacola* (CA 24), assistant R Division officer		
Feb 36–Nov 37	Ensign	*Waters* (DD 115), engineering officer		
Nov 37–Jun 38	Ensign	*Lawrence* (DD 250) communications and torpedo officer		
Jun 38–Sep 38	Lt (jg)	*Gilmer* (DD 233), engineering officer		
Oct 38–Dec 38	Lt (jg)	*Texas* (BB 35), gunnery department		
Jan 39–Jun 39	Lt (jg)	Submarine school, student		
Jun 39–Oct 39	Lt (jg)	Submarine *R-14* (SS 91), engineering officer		
Nov 39–Dec 41	Lt (jg)	*Sealion* (SS 195), engineering officer		
Dec 41	Lt (jg)	Staff ComSubPacFlt, salvage officer		
Jan 42–May 42	Lt (jg)	*Stingray* (SS 186), engineering officer		
May 42–Sep 42	Lt	*Stingray* (SS 186), executive officer		
Oct 42–Jul 42	Lt	*Lapon* (SS 260), executive officer		
Jul 42–Sep 43	Lt	*Lapon* (SS 260), executive officer	SA-3	Navy V-12 Program, University of Kansas

	ELI T. REICH		WAYNE E. MEYER	
Date	Rank	Duty	Rank	Duty
Oct 43–Feb 43	Lt Cdr	*Sealion* (SS 315), PCO	SA-3	Navy V-12 Program, University of Kansas
Mar 43–Feb 44	Lt Cdr	*Sealion* (SS 315), commanding officer	SA-3	Navy V-12 Program, University of Kansas
Mar 44–Dec 44	Cdr	*Sealion* (SS 315), commanding officer	SA-3	Navy V-12 Program, University of Kansas
Jan 45–Jul 45	Cdr	ComSubPac, asst. plans officer	SA-3	Navy V-12 Program, University of Kansas
Aug 45–Nov 45	Cdr	*Compton* (DD 705), commanding officer	SA-3	Navy V-12 Program, University of Kansas
Jan 46–Feb 46	Cdr	OpNav, member Tactical Pubs. Panel	SA-3	Navy V-12 Program, University of Kansas
Mar 46–Aug 46	Cdr	OpNav, member Tactical Pubs. Panel	Ensign	MIT, student
Sep 46–Feb 47	Cdr	U.S. Naval Academy, instructor gunnery and ordnance	Ensign	MIT, student
Feb 47–Jan 48	Cdr	U.S. Naval Academy, instructor gunnery and ordnance	Ensign	*Goodrich* (DD 831), electronics officer
Feb48–Jul 48	Cdr	Amer. Mil. Assistance Group to Turkey	Ensign	*Goodrich* (DD 831), electronics officer
Jul 48–Jul 49	Cdr	Amer. Mil. Assistance Group to Turkey	Lt (jg)	*Springfield* (CL 66), electronics officer
Jul 49–Dec 49	Cdr	Armed Forces Staff College	Lt (jg)	*Sierra* (AD 18), ASW/CIC officer
Jan 50–Feb 51	Cdr	ComSubPac staff training officer	Lt (jg)	*Sierra* (AD 18), ASW/CIC officer
Feb 51–Jul 51	Cdr	*Stoddard* (DD 566), commanding officer	Lt (jg)	*Sierra* (AD 18), ASW/CIC officer
Aug 51–Apr 52	Cdr	*Stoddard* (DD 566), commanding officer	Lt (jg)	Guided Missile School, student
Apr 52	Cdr	*Stoddard* (DD 566), commanding officer	Lt (jg)	Fleet Training Center, nuclear weapons instructor
May 52–Apr 54	Cdr	BuOrd, head of Torpedo Research Branch	Lt (jg)	Fleet Training Center, nuclear weapons instructor

	ELI T. REICH		WAYNE E. MEYER	
Date	**Rank**	**Duty**	**Rank**	**Duty**
May 54–Sep 54	Capt	BuOrd, head of Underwater Systems Division	Lt (jg)	Fleet Training Center, nuclear weapons instructor
Sep 54–May 55	Capt	BuOrd, head of Underwater Systems Division	Lt (jg)	General Line School, student
May 55–Jul 55	Capt	BuOrd, head of Underwater Systems Division	Lt	*Strickland* (DER 333), executive officer
Aug 55–Jun 56	Capt	Industrial College of the Armed Forces, student	Lt	*Strickland* (DER 333), executive officer
Jul 56–Dec 56	Capt	Submarine Squadron 8, commanding officer	Lt	*Strickland* (DER 333), executive officer
Dec 56–Aug 57	Capt	Submarine Squadron 8, commanding officer	Lt	Staff, ComDesLanFlt, asst. plans and ops officer
Aug 57–Jun 58	Capt	*Aucilla* (AO 56), commanding officer	Lt	Staff, ComDesLanFlt, asst. plans and ops officer
Jun 58–Oct 58	Capt	*Aucilla* (AO 56), commanding officer	Lt Cdr	Naval Post-Graduate School
Oct 58–Jul 60	Capt	OpNav, head of Program Management and Budget Office	Lt Cdr	Naval Post-Graduate School
Jul 60–Sep 60	Capt	OpNav, head of Program Management and Budget Office	Lt Cdr	MIT, student
Sep 60–Jun 61	Capt	*Canberra* (CAG 2), commanding officer	Lt Cdr	MIT, student
Jun 61–Dec 61	Capt	*Canberra* (CAG 2), commanding officer	Lt Cdr	*Galveston* (CLG 3), fire control/weapons officer
Feb 62–Jul 62	Capt	Asst. Chf BuWeps for Surface Missile Systems	Lt Cdr	*Galveston* (CLG 3), fire control/weapons officer
July 62–Jan 63	Capt	Head of Special Navy Task Force for Surface Missile Systems	Cdr	*Galveston* (CLG 3), fire control/weapons officer
Feb 63–Jul 63	Radm	Head of Special Navy Task Force for Surface Missile Systems	Cdr	*Galveston* (CLG 3), fire control/weapons officer

	ELI T. REICH		WAYNE E. MEYER	
Date	**Rank**	**Duty**	**Rank**	**Duty**
Jul 63–Sep 65	Radm	Director, Surface Missile Systems Project, Ofc NavMat	Cdr	BuWeps Terrier fire control manager, SMS
Sep 65–Jul 66	Radm	Commander, ASW Group 5	Cdr	Ofc NAVMAT, Project Terrier fire control manager, SMS
Jul 66–Jan 67	Radm	Commander, ASW Group 5	Capt	NAVORD, Project Terrier fire control manager
Jan 67–Feb 67	Radm	Commander, ASW Group 5	Capt	NSMSES, director of Engineering Directorate and acting exec officer
Feb 67–Jun 67	Radm	OpNav, director, Logistics Plans Division	Capt	NSMSES, director of Engineering Directorate and acting exec officer
Jun 67–May 70	Radm	OpNav, assistant director CNO for logistics	Capt	NSMSES, director of Engineering Directorate and acting exec officer
May 70–Nov 73	Vadm	Deputy assistant secretary of defense (material)	Capt	NAVORD, Chief Surface Missile System Div. (AEGIS)
Nov 73–Jul 74			Capt	NAVORD, Chief Surface Missile System Div. (AEGIS)
Jul 74–May 75			Capt	NAVSEA, project manager, Aegis/SM-2
May 75–Oct 76			Radm	NAVSEA, director for Combat Systems
Oct 76–Aug 83			Radm	NAVSEA, project manager, Aegis Shipbuilding Project
Aug 83–Nov 85			Radm	NAVSEA, deputy commander, Combat Systems

NOTES

PREFACE

1. Peter Westwick, *Stealth: The Secret Contest to Invent Invisible Aircraft* (New York: Oxford University Press, 2020), xi; Paul Kennedy, "History from the Middle," *Journal of Military History* 74, no. 1 (January 2010): 35–55. See Thomas Wildenberg, *All the Factors of Victory: Joseph Mason Reeves and the Origins of Carrier Air Power* (Annapolis, MD: Naval Institute Press, 2018).
2. Meyer's oral history left this reader with the impression that he was arrogant, a supposition supported by one long-term associate of his who revealed that Meyer keep avoiding the oral history interview, saying, "If I do this oral history, people won't need to come in and meet with me."
3. Wayne E. Meyer, *The Reminiscences of Wayne E. Meyer, Rear Adm., USN (Ret.)* [oral history transcript of 18 interviews conducted by Paul Stillwell] (Annapolis, MD: U.S. Naval Institute, 2012); Eli T. Reich, *The Reminiscences of Eli T. Reich, Vice Adm., USN (Ret.)*, 2 vols. [oral history transcript of interviews by John T. Mason Jr. (vol. 1, based on 11 interviews; vol. 2, based on 15 interviews)] (Annapolis, MD: U.S. Naval Institute, 1978).
4. James P. Rife and Rodney P. Carlisle, *The Sound of Freedom: Naval Weapons Technology at Dahlgren, Virginia 1918–2006* (Dahlgren, VA: Naval Surface Warfare Center, Dahlgren Division, 2006), 165.
5. Lawrence M. Hanser, Louis W. Miller, Herb Shukiar, and Bruce Newsome, *Developing Senior Navy Leaders: Requirements for Flag Officer Expertise Today and in the Future* (Santa Monica, CA: RAND Corporation, 2008), xv.

INTRODUCTION

1. Center for Naval Analysis, *Defense against Kamikaze Attacks in World War II and Relevance to Anti-Ship Missile Defense* (Alexandria, VA: Center for Naval Analysis, 1971), 78, table 20.
2. Delmar S. Fahrney, "History of Pilotless Aircraft and Guided Missiles," unpublished manuscript produced for the U.S. Navy, 567–69 [hereafter Fahrney, "History"], RG 72, National Archives, College Park, MD [hereafter NA]; Frederick I. Ordway, III, and Ronald C. Wakeford, *International Missile and Spacecraft Guide* (New York: McGraw-Hill, 1960), 100; Capt. Grover B. H. Hall, ACNO (Guided Missiles), Memorandum to CNO, Subj. First Successful Aircraft Interception by a Guided Missile, January 25, 1950, File VV(10), Secret Correspondence CNO Air, 1948–1951, 567–69, RG 428, Box 19, NA.
3. BuOrd [chf?] to BuAer [chf?], September 1, 1944, BuAer File F31-1, Vol. 6, as cited in Fahrney, "History," 1224–25.
4. Fahrney, "History," 1229; Walter G. Berl, "Annotated Bumblebee Initial Report," *Johns Hopkins APL Technical Digest* 3, no. 2 (1982): 171–79.
5. Audra J. Wolfe, *Competing with the Soviets: Science, Technology, and the State in Cold War America* (Johns Hopkins University Press, 2013), 29; Mathew Montoya, "Standard Missile: A

Cornerstone of Navy Theater Air Missile Defense," *Johns Hopkins APL Technical Digest* 22, no. 3 (2001): 239.

6. Fahrney, "History," 1229–30.
7. Johns Hopkins University Applied Physics Laboratory, *The First Forty Years: A Pictorial Account of the Johns Hopkins University Applied Physics Laboratory* (Baltimore, MD: Johns Hopkins University Applied Physics Laboratory, 1983), 19.

CHAPTER 1. FROM NAVAL CADET TO SUBMARINER

1. "Vice Admiral Eli T. Reich, USN (Ret.)," Navy Office of Information, Biographies Branch, June 20, 1974 (hereafter Official Navy Bio.); Eli T. Reich, *The Reminiscences of Eli T. Reich, Vice Adm., USN (Ret.)*, vol. 1 [oral history transcript of 11 interviews conducted by John T. Mason Jr.] (Annapolis, MD: U.S. Naval Institute, 1978), 3, 9 (hereafter Reich, *Reminiscences*, 1 or 2).
2. Reich, *Reminiscences*, 1:13–14.
3. *Annual Register of the United States Naval Academy* (1933, 1934, 1935), passim; Reich, *Reminiscences*, 1:17–19; Frank K. Slason, ed., *The Lucky Bag* (Annapolis, MD: Class of 1935, 1935), 220; Official Navy Bio.
4. Reich, *Reminiscences*, 1:23–24.
5. Reich, *Reminiscences*, 1:24–25.
6. Reich, *Reminiscences*, 1:28–31.
7. Reich, *Reminiscences*, 1:30, 36.
8. Reich, *Reminiscences*, 1:39.
9. Reich, *Reminiscences*, 1:40–41.
10. Reich, *Reminiscences*, 1:44–47; Norman Friedman, *U.S. Navy Destroyers: An Illustrated Design History* (Annapolis, MD: Naval Institute Press, 1982), 69; Angela D'Amico and Richard Pittenger, "A Brief History of Active Sonar," *Aquatic Mammals* 35, no. 4 (2009): 427.
11. Reich, *Reminiscences*, 1:48–49.
12. Reich, *Reminiscences*, 1:51, 57; "Why They Wear the Dolphins," *All Hands*, no. 608 (September 1967): 15.
13. Reich, *Reminiscences*, 1:52–56.
14. Reich, *Reminiscences*, 1:57–59.
15. Reich, *Reminiscences*, 1:60–61; "James Fife," *The Hall of Valor Project*, accessed July 14, 2022, https://valor.militarytimes.com/hero/27313.
16. Matthew Robert McGraw, "Beneath the Surface: American Culture and Submarine Warfare in the Twentieth Century" (master's thesis, University of Southern Mississippi, 2011), 45–46; Reich, *Reminiscences*, 1:61–62.
17. Reich, *Reminiscences*, 1:64–65; Gary E. Weir, *Building American Submarines 1914–1940* (Washington, DC: Naval Historical Center, 1991), 95.

CHAPTER 2. *SEALION* DUTY

1. Reich, *Reminiscences*, 1:67–68, 82; John D. Alden, *The Fleet Submarine in the U.S. Navy: A Design and Construction History* (Annapolis, MD: Naval Institute Press, 1979), 48, 50, 62–63, 210.
2. Reich, *Reminiscences*, 1:64, 71; Mary Lee Fowler, *Full Fathom Five: A Daughters' Search* (Tuscaloosa, AL: University Alabama Press, 2008), 176.

3. Reich, *Reminiscences*, 1:72–74.
4. Reich, *Reminiscences*, 1:66, 75; U.S. Navy, *Register of Commissioned and Warrant Officers of the United States Navy and Marine Corps: July 1, 1939* (Washington, DC: Government Printing Office, 1939), 116.
5. Reich, *Reminiscences*, 1:77–79.
6. Reich, *Reminiscences*, 1:80–81.
7. Reich, *Reminiscences*, 1:83–85; "Stuffing Box," *Marine Inbox*, accessed November 18, 2022, https://marineinbox.com/marine-exams/stuffing-box/; William Pearce, "Man Double-Acting Diesel Marine Engines," *Old Machine Press*, December 20, 2017, https://oldmachinepress.com/2017/12/20/man-double-acting-diesel-marine-engines/.
8. Reich, *Reminiscences*, 1:86–88.
9. Reich, *Reminiscences*, 1:90.
10. David L. Johnston, "A Visual Guide to the S-Class Submarines 1918–1945, Part 1: The Prototypes," *Submarine Museums* accessed August 2, 2023, https://www.submarinemuseums.org/docs/sboatsl_v2.pdf.
11. For a detailed listing, see "United States Asiatic Fleet Locations December 7, 1941," *NavSource*, accessed July 1, 2022, http://www.navsource.org/Naval/usfb.htm.
12. Reich, *Reminiscences*, 1:93–94.
13. Reich, *Reminiscences*, 1:96–98.
14. Reich, *Reminiscences*, 1:110–11.
15. Reich, *Reminiscences*, 1:113–14; John J. Domalgalski, "Disaster at Cavite," *Naval History Magazine* 32, no. 6 (December 2018), https://www.usni.org/magazines/naval-history-magazine/2018/december/disaster-cavite.
16. Reich, *Reminiscences*, 1:115–16, 119.
17. Reich, *Reminiscences*, 1:119, 129.
18. Reich, *Reminiscences*, 1:127.
19. Reich, *Reminiscences*, 1:125–27.
20. Reich, *Reminiscences*, 1:140. Reich claims that he boarded the *Stingray* at 10 or 11 o'clock on New Year's Eve. This appears to be an error, because in the Naval Historical Center's *Dictionary of American Naval Fighting Ships*, vol. 6 [hereafter *DANFS*] (Washington, DC: Government Printing Office, 1976), the entry for *Stingray* states that the submarine got underway on December 30, 1941.

CHAPTER 3. REICH'S FIRST WAR PATROLS

1. Reich, *Reminiscences*, 1:141–42.
2. Reich, *Reminiscences*, 1:145; Naval History and Heritage Command, "*Stingray* (SS 186)," *Dictionary of American Naval Fighting Ships (DANFS online)*, accessed November 17, 2022, https://www.history.navy.mil/content/history/nhhc/research/histories/ship-histories/danfs/s/stingray-ii.html; Naval Historical Center, "*Stingray* (SS 186)," *DANFS*, 6:636; Bob Hackett, "*Harbin Maru*-Class Auxiliary Hospital Ship/Transport," *CombinedFleet*, accessed July 17, 2022, http://www.combinedfleet.com/Harbin_c.htm.
3. Reich, *Reminiscences*, 1:145.
4. Reich, *Reminiscences*, 1:154–55; Paul E. Summers, as quoted in Gregory F. Michno, *USS* Pampanito *Killer-Angel* (Norman: University of Oklahoma Press, 2001). In his

Reminiscences, Reich referred to Summers as "Pete Summers," but no such person is listed in the Navy Register for 1942.

5. Thomas Wildenberg and Norman Polmar, *Ship Killers: A History of the American Torpedo* (Annapolis, MD: Naval Institute Press, 2010), 103–4.
6. Reich, *Reminiscences*, 1:150–51.
7. Reich, *Reminiscences*, 1:165.
8. Reich, *Reminiscences*, 1:151–52; U.S. Navy, *Register of Commissioned and Warrant Officers of the United States Navy and Marine Corps: July 1, 1942* (Washington, DC: Government Printing Office, 1942), 182; Naval Historical Center, "*Stingray* (SS 186)," *DANFS*, 6:636.
9. Although Reich never mentioned it in his oral history, it's likely that Reich relieved Hank Stunn when they arrived at Fremantle and that he subsequently became executive officer.
10. Reich, *Reminiscences*, 1:161–63; Naval Historical Center, "*Stingray* (SS 186)," *DANFS*, 6:636; "*Saikyo Maru* (1936–) *Saikyo Maru* (+ 1942)," *Wrecksite*, accessed July 7, 2022, https://www.wrecksite.eu/wreck.aspx?306650.
11. Reich, *Reminiscences*, 1:160–170.
12. Wildenberg and Polmar, *Ship Killers*, 132–34.
13. Brian L. Wallin, "The Torpedo Station at Newport," in *World War II Rhode Island*, by Christian McBurney, Brian L. Wallin, Patrick T. Conley, John W. Kennedy, and Maureen A. Taylor (Charleston, SC: History Press, 2017), https://www.google.com/books/edition/World_War_II_Rhode_Island/iELFDgAAQBAJ?hl=en&gbpv=1&pg=PT6&printsec=frontcover.
14. Reich, *Reminiscences*, 1:171–72. As Capt. John J. Hammerer, Jr. USN (Ret.) pointed out during his review of this manuscript ahead of its publication, the test firings are an excellent example of the importance of data taking. This is something he considers one of the fundamental principles in the development of naval weapons.
15. Reich, *Reminiscences*, 1:173–74; Reich, as quoted by Clair Blair Jr., *Silent Victory: The U.S. Submarine War against Japan* (New York: J. B. Lippincott, 1975), 402.
16. Reich, *Reminiscences*, 1:175, Blair, *Silent Victory*, 402–3.

CHAPTER 4. *SEALION* WAR PATROLS

1. Reich, *Reminiscences*, 1:177–80; Naval Historical Center, "*Lapon*," *DANFS*, 4:56.
2. Reich, *Reminiscences*, 1:181; Robert Dienesch, "Radar and the American Submarine War, 1941–1945: A Reinterpretation," *Northern Mariner/Le marin du nord* 14, no. 3 (July 2004): 32. It appears as though *Sealion II* was equipped with the earliest version of the SJ radar, which did not have a PPI display. A PPI was shortly added as a field modification identifying the radar as the model SJ-a. The SJ-1 with a PPI and an improved antenna was introduced in 1944.
3. Reich, *Reminiscences*, 1:183.
4. "Lockwood, Charles Andrew, Jr. (1890–1967)," in *The Pacific War Online Encyclopedia*, by Kent G. Budge, accessed July 29, 2022, http://pwencycl.kgbudge.com/L/o/Lockwood_Charles_A.htm.
5. Reich, *Reminiscences*, 1:185–86.
6. Reich, *Reminiscences*, 1:188.
7. Reich, *Reminiscences*, 1:189.

8. Reich, *Reminiscences*, 1:190–91.
9. Naval Historical Center, "*Sealion II*," *DANFS*, 6:416; Reich, *Reminiscences*, 1:192–93, 199.
10. Naval Historical Center, "*Sealion II*," *DANFS*, 6:416; Reich, *Reminiscences*, 1:193. The problems of the Mark 14 torpedo are beyond the scope of this work and have been extensively documented in many other publications. See, for instance, "Torpedoes That Didn't Work," in Wildenberg and Polmar, *Ship Killers*.
11. Naval Historical Center, "*Sealion II*," *DANFS*, 6:416.
12. Naval Historical Center, "*Sealion II*," *DANFS*, 6:416; Reich, *Reminiscences*, 1:194–95. It should be noted that Reich's description of the attack, thirty-four years later, differed slightly from the official version published in the *Dictionary of American Naval Fighting Ships*.
13. Naval Historical Center, "*Sealion II*," *DANFS*, 6:416–17; "IJN Minelayer *Shirataka*: Tabular Record of Movement," *CombinedFleet*, accessed August 1, 2022, http://www.combinedfleet.com/Shirataka_t.htm.
14. Reich, *Reminiscences*, 1:199, 224–26; James Marchio, "The Evolution and Relevance of Joint Intelligence Centers," *Studies In Intelligence* 48, no. 1 (2004): 43; Michno, *USS* Pampanito *Killer-Angel*, 161, 225.
15. Reich, *Reminiscences*, 1:199; Naval Historical Center, "*Sealion II*," *DANFS*, 6:417.
16. James D. Hornfischer, *Ship of Ghosts: The Story of the USS Houston, FDR's Legendary Lost Cruiser, and the Epic Saga of Her Survivors* (New York: Bantam, 2007), 348.
17. Lachlan Grant, "70th Anniversary of the Sinking of the *Rakuyo Maru*," *Australian War Memorial*, accessed August 8, 2022, https://www.awm.gov.au/articles/blog/70th-anniversary-sinking-rakuy-maru; "The Sinking of Prisoner of War Transport Ships in the Far East," *Imperial War Museums*, accessed August 3, 2022, https://www.iwm.org.uk/history/the-sinking-of; Gregory F. Michno, *Death on the Hellships: Prisoners at Sea in the Pacific War* (Annapolis, MD: Naval Institute Press, 2001), 307.
18. "The Sinking of Prisoner of War Transport Ships in the Far East"; "*Kachidoki Maru*," *POW Research Network Japan*, accessed August 8, 2022, http://www.powresearch.jp/en/archive/ship/kachidoki.html.
19. "The Sinking of Prisoner of War Transport Ships in the Far East"; National Park Service, "USS *Pampanito*," accessed August 8, 2022, https://www.nps.gov/places/uss-pampanito.htm; "*Kachidoki Maru*"; Landon L. Davis, quoted in "The Sinking of Prisoner of War Transport Ships in the Far East."
20. Reich, *Reminiscences*, 1:202–4; "*Kachidoki Maru*."

CHAPTER 5. REICH "BAGS" A BATTLESHIP

1. Naval Historical Center, "*Sealion II* (SS 315)," *DANFS*, 6:417; Reich, *Reminiscences*, 1:206–207.
2. Reich, *Reminiscences*, 1:209.
3. Reich, *Reminiscences*, 1: 211; U.S. Fleet, Headquarters of the Commander in Chief, *Radar Bulletin No. 3: Radar Operators Manual* (Washington, DC: Navy Department, April 1945), 4-SJ-9; Capt. John J. Hammerer, Jr. USN (Ret.), note to author April 4, 2023.
4. Anthony P. Tully, "The Loss of Battleship KONGO: As Told in Chapter 'November Woes' of 'Total Eclipse: The Last Battles of the IJN—Leyte to Kure 1944 to 1945,'" *CombinedFleet*, 1998, http://www.combinedfleet.com/eclipkong.html; Naval Historical Center, "*Sealion* (SS-315)," *DANFS*, 6:417; Reich, *Reminiscences*, 1:211.

5. Tully, "The Loss of Battleship KONGO."
6. Reich, *Reminiscences*, 1:211–14; Naval Historical Center, "*Sealion* (SS-315)," *DANFS*, 6:417; Tully, "The Loss of Battleship KONGO."
7. Tully, "The Loss of Battleship KONGO"; Naval Historical Center, "*Sealion* (SS-315)," *DANFS*, 6:417.
8. Reich, *Reminiscences*, 1:216; Reich, as quoted in Tully, "The Loss of Battleship KONGO."
9. Naval Historical Center, "*Sealion* (SS-315)," *DANFS*, 6:417.
10. Reich, *Reminiscences*, 1:221–24.
11. Reich, *Reminiscences*, 1:229–31.
12. Reich, *Reminiscences*, 1:235–37; E. T. Reich, CO, to SecNav, USS *Compton* (DD 705) Factual History, September 24, 1945, 3, author's collection; Naval Historical Center, "*Compton* (DD-705)," *DANFS*, 5:254.
13. Naval History and Heritage Command, "*Compton* (DD-705)," *DANFS online*, accessed July 15, 2020, https://www.history.navy.mil/research/histories/ship-histories/danfs/c/compton.html; Reich, *Reminiscences*, 1:239. Once again, Reich's memory was faulty, as the ship he identified with this incident in his oral history was the *West Virginia*.
14. Reich, *Reminiscences*, 1:241.
15. Reich, *Reminiscences*, 1:251–52; "Louis Emil Denfeld (13 April 1891–28 March 1972)," *Arlington National Cemetery*, accessed August 11, 2022, https://www.arlingtoncemetery.net/ledenfeld.htm; U.S. Navy, *Register of Commissioned and Warrant Officers of the United States Navy and Marine Corps: July 1, 1945* (Washington, DC: Government Printing Office, 1945), 5.
16. Reich, *Reminiscences*, 1:251–53.

CHAPTER 6. FROM THE NAVAL ACADEMY TO TURKEY

1. Reich, *Reminiscences*, 1:258–59, 272.
2. Reich, *Reminiscences*, 1:262, 265.
3. Reich, *Reminiscences*, 1:266–68, 273.
4. Reich, *Reminiscences*, 1:268.
5. Reich, *Reminiscences*, 1:269–70.
6. Reich, *Reminiscences*, 1:274, 278.
7. Kenneth W. Condit, *History of the Joint Chiefs of Staff: The Joint Chiefs of Staff and National Policy*, vol. 2: *1947–1949* (Office of Joint History, Office of the Chairman of the Joint Chiefs of Staff, 1996), 5, 9; "Truman Doctrine (1947)," Milestone Documents, *National Archives*, last updated February 8, 2022, https://www.archives.gov/milestone-documents/truman-doctrine; President Harry S. Truman, as quoted in "Truman Doctrine (1947)"; Tom Mackaman, "From Greece to Ukraine: 75 Years of the Truman Doctrine," *World Socialist Web Site*, June 15, 2022, https://www.wsws.org/en/articles/2022/06/16/uyno-j16.html.
8. Condit, *History of the Joint Chiefs of Staff*, 75.
9. Reich, *Reminiscences*, 1:278.
10. Reich, *Reminiscences*, 1:281–82, 284.
11. Reich, *Reminiscences*, 1:285–88, U.S. Navy, Office of the Chief of Naval Operations, *Catalogue of Advanced Base Functional Components*, 3rd ed. (Washington, DC: Office of the Chief Naval Officer, 1945), 128. The N2C camp was the same as N1C, but reduced for 100 men.

12. Reich, *Reminiscences*, 1:287–89; for the J5B Torpedo Depot (medium), see U.S. Navy, Office of the Chief of Naval Operations, *Catalogue of Advanced Base Functional Components*, 112 (see page 118 for J11A Mine Assembly Depot).
13. Reich, *Reminiscences*, 1:290–91.
14. Reich, *Reminiscences*, 1:294–95.
15. Reich, *Reminiscences*, 1:296–98, 301–3.
16. Reich, *Reminiscences*, 1:304–5.
17. Reich, *Reminiscences*, 1:306–7.
18. Reich, *Reminiscences*, 1:299.
19. Reich, *Reminiscences*, 1:314–15.
20. Reich, *Reminiscences*, 1:316.
21. Reich, *Reminiscences*, 1:318.
22. Reich, *Reminiscences*, 1:322–23.
23. Reich, *Reminiscences*, 1:322–25.
24. Reich, *Reminiscences*, 1:326–27.
25. Reich, *Reminiscences*, 1:327–29.
26. Reich, *Reminiscences*, 1:330, 334–35.

CHAPTER 7. FROM STAFF OFFICER TO DESTROYER COMMAND

1. Armed Forces Staff College, *Command History of the Armed Forces Staff College 1946–1981* (Washington, DC.: Armed Forces Staff College, 1983[?]), 33.
2. Armed Forces Staff College, *Command History*, 13; Reich, *Reminiscences*, 1:340.
3. Reich, *Reminiscences*, 1:342; Norman Polmar and K. J. Moore, *Cold War Submarines: The Design and Construction of U.S. and Soviet Submarine* (Dulles, VA: Potomac Books, 2004), 20; Gary E. Weir, *Forged in War: The Naval-Industrial Complex and American Submarine Construction, 1940–1961* (Washington, DC: Naval Historical Center, 1993), 134; "U.S. Navy Hunter-Killer Submarines," *Weapons and Warfare*, accessed August 30, 2022, https://weaponsandwarfare.com/2017/01/25/us-navy-hunter-killer-submarines/; Mike Smolinski, "K-1 Barracuda (SSK-1) (SST-3)," *NavSource Submarine Photo Archive*, accessed August 30, 2022, http://www.navsource.org/archives/08/08550.htm; "Sonar Receiving Set AN/BQR-4 (Preliminary)," November 9, 1951, in U.S. Navy, Naval Ship Systems Command, *Catalogue of Electronic Equipment, NavShips 900,116*, Suppl. 5, December 1952, 23.
4. Reich, *Reminiscences*, 1:343.
5. "U.S. Navy Hunter-Killer Submarines"; Weir, *Forged in War*, 134.
6. Reich, *Reminiscences*, 1:345–46.
7. Reich, *Reminiscences*, 1:347–48.
8. Reich, *Reminiscences*, 1:348–49.
9. Reich, *Reminiscences*, 1:349–50.
10. Reich, *Reminiscences*, 1:350–51, 355.
11. Reich, *Reminiscences*, 1:351–52.
12. Reich, *Reminiscences*, 1:352.
13. Reich, *Reminiscences*, 1:356–58.
14. Reich, *Reminiscences*, 1:367.
15. Reich, *Reminiscences*, 1:358–59, 366.
16. Reich, *Reminiscences*, 1:359–60.

17. Reich, *Reminiscences*, 1:360–363.
18. Reich, *Reminiscences*, 1:363–34.
19. Reich, *Reminiscences*, 1:358, 364–64.
20. This observation was made by Capt. John H. Hammerer Jr., USN (Ret.).
21. Reich, *Reminiscences*, 1:369–70; Joseph W. C. Harpster, "USS *Stoddard* DD566 in the Korean War Era," *USSStoddard*, accessed September 3, 2022, http://www.ussstoddard.org/koreaharpster.html.
22. Reich, *Reminiscences*, 1:373; Harpster, "USS *Stoddard* DD566 in the Korean War Era"; "3"/50 (7.62 cm) Marks 27, 33 and 34," *NavWeaps*, accessed September 4, 2021, http://www.navweaps.com/Weapons/WNUS_3-50_mk27-33-34.php; U.S. Navy, Bureau of Ordnance, *Explosive Ordnance* (Washington, DC: Bureau of Ordnance, 1947), 26, 70.
23. Reich, *Reminiscences*, 1:273–74.

CHAPTER 8. FROM THE TORPEDO RESEARCH BRANCH TO THE INDUSTRIAL WAR COLLEGE

1. Reich, *Reminiscences*, 1:375–76, 379–80.
2. Reich, *Reminiscences*, 1: 377.
3. Reich, *Reminiscences*, 1:381.
4. Reich, *Reminiscences*, 1:382–83.
5. E. W. Jolie, *A Brief History of U.S. Navy Torpedo Development* (Newport, RI: Naval Underwater Systems Center, 1978), 105, 115; Reich, *Reminiscences*, 1:391.
6. Reich, *Reminiscences*, 1:386.
7. Reich, *Reminiscences*, 1:386; Wildenberg and Polmar, *Ship Killers*, 147; National Defense Research Committee, *Acoustic Torpedoes: Summary Technical Report of Division 6, NDRC* (Washington, DC: Office of Scientific Research and Development, 1946), 31, 78.
8. Reich, *Reminiscences*, 1:386; Jolie, *A Brief History of U.S. Navy Torpedo Development*, 105.
9. Reich, *Reminiscences*, 1:386; John Merrill and Lionel D. Wyld, *Meeting the Submarine Challenge: A Short History of the Naval Underwater Systems Center* (Washington, DC: Government Printing Office, 1997), 143; Wildenberg and Polmar, *Ship Killers*, 153.
10. Reich, *Reminiscences*, 1:389–90.
11. Reich, *Reminiscences*, 1:390.
12. Reich, *Reminiscences*, 1:406–9; Michno, *USS* Pampanito *Killer-Angel*, 43.
13. Reich, *Reminiscences*, 1:394–95
14. Reich, *Reminiscences*, 1:396–97.
15. Reich, *Reminiscences*, 1:398–399.
16. Reich, *Reminiscences*, 1:400, 410.
17. Mary Kanagy, ed., "BuOrd Relays CNO Praise for NOL Part in Special Job," *Report . . . U.S. Naval Ordnance Laboratory* 11, no. 10 (April 1955): 1.

CHAPTER 9. FROM STUDENT TO BUDGET OFFICER, WITH SEA COMMANDS BETWEEN

1. Theodore W. Bauer, *History of the Industrial College of the Armed Forces* (Washington, DC: Alumni Association of the Industrial College of the Armed Forces, 1983), iv–3.
2. Reich, *Reminiscences*, 1:415, 420.
3. Reich, *Reminiscences*, 1:416–18.

4. Reich, *Reminiscences*, 1:421–22.
5. Reich, *Reminiscences*, 1:424–25.
6. Reich, *Reminiscences*, 1:425–26; Naval History and Heritage Command, "*Tringa*," *DANFS online*, accessed September 21, 2022, https://www.history.navy.mil/research/histories/ship-histories/danfs/t/tringa.html.
7. Naval History and Heritage Command, "*Aucilla* (AO-56)," *DANFS online*, accessed September 21, 2022, https://www.history.navy.mil/research/histories/ship-histories/danfs/a/aucilla.html; Reich, *Reminiscences*, 1:427, 431.
8. Reich, *Reminiscences*, 1:433–44.
9. Reich, *Reminiscences*, 1:437–38.
10. Reich, *Reminiscences*, 1:440–41.
11. Reich, *Reminiscences*, 1:440–41.
12. "John T. Hayward." *DBpedia*, accessed September 22, 2022, https://dbpedia.org/page/John_T._Hayward; John T. Hayward, "History of the U.S. Naval War College: Reminiscences of NWC Presidency, 1966–68 and U.S. Navy Career," interview by Dr. Evelyn M. Cherpak, May 1977, Naval War College, 21, https://archive.org/details/oralhistorytypes00hayw/page/n7/mode/2up.
13. Reich, *Reminiscences*, 1:441–42.
14. Reich, *Reminiscences*, 1:445–48.
15. Reich, *Reminiscences*, 1:448–49, 451.
16. Reich, *Reminiscences*, 1:452.
17. Reich, *Reminiscences*, 1:453.
18. Reich, *Reminiscences*, 1:455.
19. Reich, *Reminiscences*, 1:459.

CHAPTER 10. COMMANDING A GUIDED MISSILE CRUISER

1. Naval Historical Center, *Canberra* I (CA-70), *DANFS*, 2:23–24; Norman Friedman, *U.S. Cruisers: An Illustrated Design History* (Annapolis, MD: Naval Institute Press, 1984), 380; Norman Friedman, *Naval Radar* (Annapolis, MD: Naval Institute Press, 1981), 57, 159, 181.
2. Chris Chant, "First of the Great Naval SAMs—the RIM-2 Terrier," *Chris Chant's Blog*, accessed December 2, 2022, https://cmchant.com/the-rim-2-terrier-the-first-of-the-great-naval-sams/; Andreas Parsch, "RIM-2," *Directory of U.S. Military Rockets and Missiles*, 2004, http://www.designation-systems.net/dusrm/m-2.html; Reich, *Reminiscences*, 1:470.
3. Naval Historical Center, *Canberra* I (CA-70), *DANFS*, 2:23–24; Reich, *Reminiscences*, 1:461, 465–66.
4. Reich, *Reminiscences*, 1:461, 63.
5. Reich, *Reminiscences*, 1:464.
6. Reich, *Reminiscences*, 1:465–66.
7. Reich, *Reminiscences*, 1:467.
8. Reich, *Reminiscences*, 1:468.
9. Reich, *Reminiscences*, 1:469–70.
10. Reich, *Reminiscences*, 1:471–72.
11. Reich, *Reminiscences*, 1:474–75.
12. Reich, *Reminiscences*, 1:445–46.
13. Reich, *Reminiscences*, 1:477–78.

14. Reich, *Reminiscences*, 1:481–82.
15. Reich, *Reminiscences*, 1:484.
16. Reich, *Reminiscences*, 1:486.
17. Reich, *Reminiscences*, 1:488.
18. Reich, *Reminiscences*, 1:489–90.
19. Reich, *Reminiscences*, 1:498–99, 500.
20. Reich, *Reminiscences*, 1:492.

CHAPTER 11. ELI REICH, GUIDED MISSILE CZAR

1. Reich, *Reminiscences*, 1:492–93; U.S. Navy, Bureau of Personnel, *Register of the Commission and Warrant Officers of the Navy of the United States, Including Officers of the Marine Corps* (Washington, DC: Secretary of the Navy, 1961), 9.
2. Reich, *Reminiscences*, 1:493–94.
3. Reich, *Reminiscences*, 1:494–95.
4. Reich, *Reminiscences*, 1:495–96.
5. Reich, *Reminiscences*, 1:500–501.
6. Reich, *Reminiscences*, 1:501–2.
7. Thomas W. Sheppard, as quoted in William K. Klingaman, *APL—Fifty Years of Service to the Nation* (Laurel, MD: Johns Hopkins University Applied Physics Laboratory, 1993), 146–47; Malcolm Muir Jr., *Black Shoes and Blue Water* (Washington, DC: Naval Historical Center, 1996), 125.
8. Klingaman, *APL—Fifty Years of Service to the Nation*, 148; Muir, *Black Shoes and Blue Water*, 139; Reich, *Reminiscences*, 1:507.
9. Reich, *Reminiscences*, 1, 507–9.
10. Reich, *Reminiscences*, 1:509.
11. Reich, *Reminiscences*, 1:510–12.
12. Reich, *Reminiscences*, 1:514–16.
13. Reich, *Reminiscences*, 1:516–21.
14. Reich, *Reminiscences*, 2:534, 541–42.
15. Reich, *Reminiscences*, 2:543–45.
16. Reich, *Reminiscences*, 2:546.
17. Reich, *Reminiscences*, 2:547–48; David L. Boslaugh, *When Computers Went to Sea: The Digitization of the United States Navy* (Los Alamitos, CA: IEEE Computer Society, 1999), 316.
18. Reich, *Reminiscences*, 2:565.
19. Boslaugh, *When Computers Went to Sea*, 321; Friedman, *Naval Radar*, 165.
20. Boslaugh, *When Computers Went to Sea*, 322; David L. Boslaugh, "First-Hand: Moving the Firing Key to NTDS—Chapter 6 of the Story of the Naval Tactical Data System," *ETHW* (wiki), last modified May 12, 2021, https://ethw.org/First-Hand:Moving_the_Firing_Key_to_NTDS_Chapter_6_of_the_Story_of_the_Naval_Tactical_Data_System.
21. Boslaugh, "First-Hand: Moving the Firing Key to NTDS."
22. Reich, *Reminiscences*, 1:478; Reich, *Reminiscences*, 2:597–599; Walter S. Poole, *Adapting to Flexible Response 1960–1968: History of Acquisition in the Department of Defense, Volume II* (Washington, DC: Historical Office, Office of the Secretary of Defense, 2013), 313–14; Marion E. Oliver, "Terrier/Tartar: Pacing the Threat," *Johns Hopkins APL Technical Digest* 2, no. 4 (1981): 257.

23. Wayne E. Meyer, *The Reminiscences of Wayne E. Meyer, Rear Adm., USN (Ret.)* [oral history transcript of 18 interviews conducted by Paul Stillwell] (Annapolis, MD: U.S. Naval Institute, 2012), 269, 312–13 [hereafter Meyer, *Reminiscences*].

CHAPTER 12. THE V-12 PROGRAM

1. Meyer, *Reminiscences*, 25, passim; "Wayne E. Meyer," *Chariton County Historical Society*, accessed July 23, 2022, https://m.facebook.com/charitoncountymuseum/photos/wayne-e-meyerwayne-e-meyer-was-born-in-brunswick-missouri-on-april-21-1926-to-eu/1792752124316360/.
2. Henry C. Herge, *Navy V-12* (Paducah, KY: Turner Publishing, 1996), 23; Julius A. Furer, *Administrative History of Navy Department in World War II* (Washington, DC: Department of the Navy, 1959), 276; Meyer, *Reminiscences*, 53, 56, 62.
3. Meyer, *Reminiscences*, 62–66.
4. Meyer, *Reminiscences*, 66–68; Furer, *Administrative History* , 276.
5. U.S. Navy, Bureau of Naval Personnel, *The Navy College Training Program V-12: Curricula Schedules/Course Descriptions, U.S. Navy* (1943), 6–7, https://www.google.com/books/edition/The_Navy_College_Training_Program_V_12/AilDwgEACAAJ?hl=en&gbpv=1&printsec=frontcover; Meyer, *Reminiscences*, 69–70.
6. Meyer, *Reminiscences*, 71–73; U.S. Navy, Bureau of Naval Personnel, *The Navy College Training Program V-12*, 7.
7. Meyer, *Reminiscences*, 86–87.
8. Meyer, *Reminiscences*, 71–73; U.S. Navy, Bureau of Naval Personnel, *The Navy College Training Program V-12*, 7.
9. Meyer, *Reminiscences*, 74–75.
10. Meyer, *Reminiscences*, 75–76.

CHAPTER 13. FROM MIT TO DUTY AT SEA

1. Meyer, *Reminiscences*, 77–78.
2. Meyer, *Reminiscences*, 78–79. When Meyer stated that he was sent to the Army Air Forces base in Oakland, his memory must have failed because the only air base in Oakland was run by the Navy.
3. Meyer, *Reminiscences*, 81.
4. Meyer, *Reminiscences*, 81–82, 96.
5. Meyer, *Reminiscences*, 81; Friedman, *U.S. Destroyers*, 226; "Some Notes on the SP," *BuShips Electron* 1, no. 3 (September 1945): 16.
6. Meyer, *Reminiscences*, 92.
7. Meyer, *Reminiscences*, 93.
8. Meyer, *Reminiscences*, 93–95.
9. Meyer, *Reminiscences*, 95–96, 102, 195.
10. Meyer, *Reminiscences*, 93, 97–98.
11. Meyer, *Reminiscences*, 100.
12. Meyer, *Reminiscences*, 99; David Mindell, *Between Human and Machine: Feedback, Control, and Computing before Cybernetics* (Baltimore, MD: Johns Hopkins University Press, 2002), 55, 63; Meyer, *Reminiscences*, 93.
13. Meyer, *Reminiscences*, 105–6.

14. Meyer, *Reminiscences*, 107–8.
15. Naval Historical Center, "*Goodrich* (DD 831)," *DANFS*, 3:120; Meyer, *Reminiscences*, 110, 115.
16. Meyer, *Reminiscences*,117.
17. Meyer, *Reminiscences*, 121, 124–25.
18. Meyer, *Reminiscences*, 120, 125.

CHAPTER 14. FROM ELECTRONICS OFFICER TO NUCLEAR WEAPONS INSTRUCTOR

1. Meyer, *Reminiscences*, 127–28.
2. Meyer, *Reminiscences*, 128–29.
3. Meyer, *Reminiscences*, 129–30; Friedman, *Naval Radar*, 146, 156; "SC Family of Radars," *Radartutorial*, accessed August, 26, 2022, https://www.radartutorial.eu/19.kartei/11.ancient/karte011.en.html. Meyer seems to be confused here with a later radar. Although Friedman lists a number of different SPS radars, none was available in 1948. The "bedspring" antenna was very characteristic of the SC radars deployed in World War II.
4. Meyer, *Reminiscences*, 129–30.
5. Meyer, *Reminiscences*, 132–34.
6. Meyer, *Reminiscences*, 132, 149.
7. Naval Historical Center, *Sierra* (AD-18), *DANFS*, 6:449–501; Meyer, *Reminiscences*, 125–50. Meyer was promoted to lieutenant junior grade on February 2, 1949; U.S. Navy, *Register of Commissioned and Warrant Officers of the United States Navy and Marine Corps: July 1, 1944* (Washington, DC: Government Printing Office, 1944).
8. Naval Historical Center, *Sierra* (AD-18), *DANFS*, 6:499–501; Meyer, *Reminiscences*, 153–54.
9. Naval Historical Center, *Sierra* (AD-18), *DANFS*, 6:499–501; Meyer, *Reminiscences*, 153–54, 158.
10. Meyer, *Reminiscences*, 159.
11. Meyer, *Reminiscences*, 169–61.
12. Jessica Gressett, ed., *Army Air Artillery Defense School* (Fort Bliss, TX: U.S. Army, n.d.), accessed November 25, 2022, https://srmsc.org/pdf/005240p0.pdf.
13. Meyer, *Reminiscences*, 162–63; Andreas Parsch, "SAM-N-2/SAM-N-2 [Lark]," *Directory of U.S. Military Rockets and Missiles*, 2004, https://www.designation-systems.net/dusrm/appl/sam-n-2.html; Eric Burgess, *Guided Weapons* (New York: Macmillan, 1957), 143–44; G. B. H. Hall, ACNO (Guided Missiles), to CNO, January 5, 1950, File VV(10), Secret Correspondence DCNO Air (OP02-OP5), 1948–1951, RG 425, Box 19, National Archives, College Park, MD.
14. Meyer, *Reminiscences*, 162, 168–69; Ralph E. Gibson, "Alexander Kossiakoff: His Life and Career, Part One." *Johns Hopkins APL Technical Digest* 27, no. 1 (2006): 6; Walter G. Berl, "Annotated Bumblebee Initial Report," *Johns Hopkins APL Technical Digest* 3, no. 2 (1982): 173–74, https://secwww.jhuapl.edu/techdigest/Content/techdigest/pdf/V03-N02/03-02-Berl.pdf.
15. Meyer, *Reminiscences*, 184.
16. Meyer, *Reminiscences*, 184; Norman Polmar and Robert S. Norris, *The U.S. Nuclear Arsenal: A History of Weapons and Delivery Systems since 1945* (Annapolis, MD: Naval Institute Press, 2009), 38, 48–49.
17. Meyer, *Reminiscences*, 184–86, 192.

CHAPTER 15. GRADUATE SCHOOLS AND MORE SEA DUTY

1. U.S. Navy, Bureau of Naval Personnel, "The General Line School Will Offer Further Professional Education," *Bureau of Personnel Training Bulletin*, March 15, 1946, 2–3.
2. Meyer, *Reminiscences*, 188–89.
3. Friedman, *U.S. Destroyers*, 230–31; Friedman, *Naval Radar*, 157–58; "AN/SPS-4," *Radartutorial*, accessed September 12, 2022, https://www.radartutorial.eu/19.kartei/11.ancient2/karte025.en.html.
4. Meyer, *Reminiscences*, 205; Collins Radio Co., *HF Communication Equipment* [catalog] *1984–85*, 15. Meyer must have been referring to the number of cabinets, which he remembered incorrectly.
5. Meyer, *Reminiscences*, 203; Friedman, *U.S. Destroyers*, 231.
6. Meyer, *Reminiscences*, 197–98, 203.
7. Meyer, *Reminiscences*, 209.
8. Meyer, *Reminiscences*, 210.
9. Meyer, *Reminiscences*, 209–10.
10. Meyer, *Reminiscences*, 215–16.
11. Meyer, *Reminiscences*, 217.
12. Meyer, *Reminiscences*, 221–22.
13. Meyer, *Reminiscences*, 225.
14. Meyer, *Reminiscences*, 226–31.

CHAPTER 16. TALOS FIRE CONTROL OFFICER

1. Meyer, *Reminiscences*, 240, 242.
2. Elmer G. Robinson, "The Talos Ship System," *Johns Hopkins APL Technical Digest* 3, no. 2 (1982): 162–63; Friedman, *U.S. Cruisers*, 382, 385; Norman Friedman, *U.S. Naval Weapons: Every Gun, Missile, Mine, and Torpedo Used by the U.S. Navy from 1883 to the Present Day* (Annapolis, MD: Naval Institute Press, 1988), 245, 273.
3. Robinson, "The Talos Ship System," 165.
4. L. L. Frank, in a lecture at the Naval War College, as quoted by Muir, *Black Shoes and Blue Water*, 65.
5. Friedman, *U.S. Naval Weapons*, 151.The speed of the Talos 6b is derived from the 2,000 feet per second values listed in William Garten Jr. and Frank A. Dean, "Evolution of the Talos Missile," *Johns Hopkins APL Technical Digest* 3, no. 2 (1982): 120; Fletcher C. Paddison, "The Talos Control System," *Johns Hopkins APL Technical Digest* 3, no. 2 (1982): 154–55.
6. Joseph Gulick, W. Colleman Hyatt, and Oscar M. Martin, Jr. "The Talos Guidance System," *Johns Hopkins APL Technical Digest* 3, no. 2 (1982): 143.
7. Naval Historical Center, "*Galveston* (CL 93)," *DANFS*, 3:14; for information on the SPW-1, see Friedman, *Naval Radar*, 182.
8. Robinson, "The Talos Ship System," 163; Phillip R. Hays, "USS *Oklahoma City* CL91/CLG5/CG5 SPG-49 Talos Mk 77 Guided Missile Fire Control System," *Okieboat*, accessed December 3, 2022, https://www.okieboat.com/Talos%20fire%20control%20system.html.
9. Meyer, *Reminiscences*, 259.
10. Meyer, *Reminiscences*, 260.

11. Hays, "USS *Oklahoma City*"; Tyler Rogoway, "The Talos Missile Had a Wonderfully Complex Shipboard Assembly Line of a Launch System," *The War Zone*, August 2, 2008, https://www.thedrive.com/the-war-zone/22599/the-talos-missile-had-a-wonderfully-complex-shipboard-assembly-line-of-a-launch-system.
12. Andreas Parsch, "RIM-8," *Directory of U.S. Military Rockets and Missiles*, https://www.designation-systems.net/dusrm/m-8.html; Garten and Dean, "The Talos Control System," 121–22; Gulick, Hyatt, and Martin, "The Talos Guidance System," 148.
13. Meyer, *Reminiscences*, 262.
14. Meyer, *Reminiscences*, 309, 311, 350.

CHAPTER 17. ELI REICH, DIRECTOR, ADVANCED SURFACE MISSILE SYSTEM

1. Boslaugh, *When Computers Went to Sea*, 316.
2. Boslaugh, *When Computers Went to Sea*, 316; Reich, *Reminiscences*, 2: 572–75; Robert K. Irvine, "Surface Missile Systems Management Action Keeps Pace with Accelerating Technology, *Navy Management Review* 12, no. 8 (August 1967): 30; Russ Pyle, "Countdown in Camden," *Surface Warfare Magazine* 3, no. 2 (February 1978): 16–19.
3. Testimony before Congress of Robert W. Morse, Assistant Secretary of the Navy for R&D, March 9, 1965, in *Hearings before and Special Reports Made by Committee on Armed Services of the House of Representatives on Subjects Affecting the Naval and Military Establishments, 1965*, 89th Cong., 1st Sess. (Washington, DC: Government Printing Office, 1965), 1000–1001.
4. Reich, *Reminiscences*, 2:534; Milton Gussow and Edward C. Prettyman, "Typhon: A Weapon System ahead of Its Time," *Johns Hopkins APL Technical Digest* 13, no. 1 (1992): 82; Klingaman, *APL—Fifty Years of Service to the Nation*, 141–42.
5. Reich, *Reminiscences*, 2:534, 580–82.
6. Reich, *Reminiscences*, 2:583; Friedman, *U.S. Naval Weapons*, 155; Friedman, *U.S. Destroyers*, 224, 339; testimony before Congress of Secretary of Defense Robert S. McNamara, February 23, 1965, in *Hearings before and Special Reports Made by Committee on Armed Services . . . 1965*, 89th Cong., 1st Sess., 533.
7. Muir, *Black Shoes and Blue Water*, 141–42; George F. Emch, "Air Defense for the Fleet," *Johns Hopkins APL Technical Digest* 13, no. 1 (1992): 48–49; Reich, *Reminiscences*, 2:588.
8. Testimony before Congress of Robert W. Morse, Assistant Secretary of the Navy for R&D, March 9, 1965, in *Hearings before and Special Reports Made by Committee on Armed Services . . . 1965*, 89th Cong., 1st Sess., 1042; Norman Friedman, "Missiles in Fleet Air Defense," unpublished history prepared for the U.S. Navy, 199, RG 72, National Archives, College Park, MD.
9. Reich, *Reminiscences*, 2:602; Kevin Bash and Angelique Bash, *A Brief History of NORCO* (Charleston, SC: History Press, 2013), 183; Dennis Casebier, as quoted in Christopher Okula, "Navy Celebrates 50 Years of Truth, Commemorates Golden Adversary of Independent Assessment," *U.S. Navy*, accessed December 2, 2022, https://www.navy.mil/submit/display.asp?story_id=80607.
10. Reich, *Reminiscences*, 2:601–603.
11. U.S. Navy, Naval Sea Systems Command, "Warfare Centers NSWC Corona Division," *NAVSEA*, accessed November 15, 2018, http://www.navsea.navy.mil/Home/Warfare-Centers/NSWC-Corona/Who-We-Are/Command-History.

12. Reich, *Reminiscences*, 2:604.
13. Reich, *Reminiscences*, 2:588; Friedman, "Missiles in Fleet Air Defense," 283.
14. Reich, *Reminiscences*, 2:588–89; Friedman, "Missiles in Fleet Air Defense," 283. Neither author is clear on which contractors were selected.
15. Reich, *Reminiscences*, 2:591–92; Bruce D. Inman, "From Typhon to Aegis—The Issues and Their Resolution," *Naval Engineers Journal* 100, no. 3 (May 1988): 65–66.
16. Marion E. Oliver and William N. Sweet, "Standard Missile: The Common Denominator," *Johns Hopkins APL Technical Digest* 2, no. 4 (1981): 284; James D. Flanagan and William N. Sweet, "Aegis: Advanced Surface Missile System," *Johns Hopkins APL Technical Digest* 2, no. 4 (1981): 244; Boslaugh, *When Computers Went to Sea*, 281–83; Kenneth P. Werrell, *Archie to SAM: A Short Operational History of Ground-Based Air Defense* (Maxwell Air Force Base, AL: Air University Press, 2005), 202; Inman, "From Typhon to Aegis," 71; Friedman, "Missiles in Fleet Air Defense," 294–95.
17. Reich, *Reminiscences*, 2:606.
18. Reich, *Reminiscences*, 2:605, 607; Boslaugh, *When Computers Went to Sea,* 146.
19. Robert C. Manke, "Overview of U.S. Navy Antisubmarine Warfare (ASW) Organization during the Cold War Era," NUWC-NPT Technical Report 11,890 (Newport, RI: Naval Undersea Warfare Center Division, 2008), 15.
20. Reich, *Reminiscences*, 2:608–9.

CHAPTER 18. ELI REICH'S LAST SEA DUTY

1. Reich, *Reminiscences*, 2:613, 616.
2. Reich, *Reminiscences*, 2:617–18.
3. Reich, *Reminiscences*, 2:619.
4. Reich, *Reminiscences*, 2:622–24; "TRW Systems, Operating Group of TRW, Inc.," *Signal*, September 1966, 83.
5. Reich, *Reminiscences*, 2:629.
6. Reich, *Reminiscences*, 2:631; "USS Kearsarge (CVS 33) WestPac Cruise Book 1966," *Navy*, accessed October 19, 2022, https://www.navysite.de/cruisebooks/cv33-66/index.html.
7. Reich, *Reminiscences*, 2:631–22.
8. Reich, *Reminiscences*, 2:632–33.
9. Reich, *Reminiscences*, 2:633–35.
10. Reich, *Reminiscences*, 2:635–36.
11. Reich, *Reminiscences*, 2:636–37.
12. Reich, *Reminiscences*, 2:637–38.
13. Reich, *Reminiscences*, 2:639, 643–44.
14. Reich, *Reminiscences*, 2:639, 644–45; Jolie, *A Brief History of U.S. Navy Torpedo Development*, 116.
15. Reich, *Reminiscences*, 2:640.
16. Peter Fey, *Bloody Sixteen: The USS Oriskany and Air Wing 16 during the Vietnam War* (Lincoln, NE: Potomac Books 2018), 145–46.
17. Fey, *Bloody Sixteen*, 147; Reich, *Reminiscences*, 2:641.
18. Reich, *Reminiscences*, 2:648, 651.

CHAPTER 19. ELI REICH, ACCOUNTING GURU

1. Reich, *Reminiscences*, 2:652–53; Eli Thomas Reich, "Transcript of Naval Service," National Military Records Center, National Archives, St. Louis, MO.
2. Reich, *Reminiscences*, 2:656–57.
3. Reich, *Reminiscences*, 2:657–58.
4. Reich, *Reminiscences*, 2:663–64.
5. Reich, *Reminiscences*, 2:664–65.
6. Reich, *Reminiscences*, 2:665–67; U.S. House of Representatives, *Hearings before the Subcommittee of the Committee on Appropriations House of Representatives*, 86th Cong., 2nd Sess., Department of the Navy (Washington, DC: Government Printing Office, 1958), 650.
7. Reich, *Reminiscences*, 2:672, 683.
8. Reich, *Reminiscences*, 2:674.
9. Reich, *Reminiscences*, 2:675–76.
10. Reich, *Reminiscences*, 2:676–77.
11. Reich, *Reminiscences*, 2:677–68.
12. Reich, *Reminiscences*, 2:684.
13. Reich, *Reminiscences*, 2:684–85.
14. Reich, *Reminiscences*, 2:686, 700–702.
15. Joe Ferrara, "DoD's 5000 Documents: Evolution and Change in Defense Acquisition Policy," *Acquisition Review Quarterly* (Fall 1996): 111, https://www.dau.edu/library/arj/ARJ/arq96/ferrar.pdf, 111; Reich, *Reminiscences*, 2:702.
16. Reich, *Reminiscences*, 2:704–5.
17. Reich, *Reminiscences*, 2:750–51, 764.
18. U.S. House of Representatives, *Department of Defense Appropriations for 1978 Hearings before a Subcommittee of the Committee on Appropriations House of Representatives*, 95th Cong., 1st Sess., Ingalls Shipyard/Litton Industries (Washington, DC: Government Printing Office, 1977), 316; John H. Rubel, *Memoirs III: Time and Chance 1959–1976* (Santa Fe, NM: Kay Say Publications, 2006), 297; Reich, *Reminiscences*, 2:756; Kenneth G. Cooper, "Naval Ship Production: A Claim Settled and a Framework Built," *Interfaces* 10, no. 6 (December 1980): 22.
19. Reich, *Reminiscences*, 2:760.
20. Reich, *Reminiscences*, 2:758–61.
21. Reich, *Reminiscences*, 2:764.

CHAPTER 20. WAYNE MEYER HONES THE SKILLS OF A MISSILEER

1. William W. Berry, "Radars and Computers aboard USS *Sterett* and Other Warships during the Vietnam War," *NavWeaps*, accessed September 30, 2022, http://www.navweaps.com/index_tech/tech-088.php; William A. Cockell Jr., "The DLG Modernization Program," U.S. Naval Institute *Proceedings* 98, no. 1 (January 1972): 101–2.
2. Boslaugh, *When Computers Went to Sea*, 119–21.
3. Boslaugh, *When Computers Went to Sea*, 302.
4. Wayne E. Meyer, "Our Navy—Like Our Lives—Is Continuous," *Johns Hopkins APL Technical Digest* 23, no. 2 and 3 (2002): 111; Meyer, *Reminiscences*, 296.
5. Meyer, *Reminiscences*, 296–97; Berry, "Radars and Computers aboard USS *Sterett* and Other Warships during the Vietnam War."

6. Meyer, *Reminiscences*, 298–99.
7. Boslaugh, *When Computers Went to Sea*, 324; Boslaugh, "First-Hand: Moving the Firing Key to NTDS."
8. Boslaugh, "First-Hand: Moving the Firing Key to NTDS."
9. Cockell, "The DLG Modernization Program," 102.
10. U.S. Navy, "USS Leahy (DLG-16/CG-16)" [photograph], *Naval History and Heritage Command*, accessed September 30, 2022, https://www.history.navy.mil/content/history/nhhc/our-collections/photography/us-navy-ships/alphabetical-listing/l/uss-leahy—dlg-16-cg-16–0.html; Meyer, *Reminiscences*, 338–43.
11. Meyer, *Reminiscences*, 300–301.
12. Cockell, "The DLG Modernization Program," 102.

CHAPTER 21. GENESIS OF AEGIS

1. Friedman, "Missiles in Fleet Air Defense," 295; Reich, *Reminiscences*, 2:592.
2. Friedman, "Missiles in Fleet Air Defense," 295; Oliver and Sweet, "Standard Missile: The Common Denominator"; John S. Foster Jr., "Oral History Interview I," December 3, 1968, by Dorothy Pierce LBJ Presidential Library, Austin, TX, 3; Meyer, *Reminiscences*, 471; Rife and Carlisle, *The Sound of Freedom*, 166; U.S. Senate, *Fiscal Year 1975 Authorization for Military Procurement, Research and Development, and Active Duty, Selected Reserve, and Civilian Personnel Strengths: Hearings before the Committee on Armed Services, United States Senate*, 93rd Cong., 2nd Sess., on S.3000 (Washington, DC: Government Printing Office, 1974), 4877.
3. Joseph T. Threston, "The AEGIS Weapon System," *Naval Engineers Journal* 121, no. 3 (October 5, 2009): 87.
4. Oliver and Sweet, "Standard Missile: The Common Denominator," 284; Ralph E. Hawes and James R. Whalen, "The Standard Missile/Aegis Story," *Naval Engineers Journal* 121. no. 3 (October 5, 2009):139; Friedman, "Missiles in Fleet Air Defense," 203–5.
5. Emch, "Air Defense for the Fleet," 48; Friedman, "Missiles in Fleet Air Defense," 295.
6. Oliver and Sweet, "Standard Missile: The Common Denominator," 284; Emch, "Air Defense for the Fleet," 50.
7. Wilbert J. Nace, "Term Project Report Systems Acquisition Management" (Thesis, U.S. Naval Post Graduate School, Monterey, CA, 1972), 45; John F. Morton, "Profound Simplicity, Part I: Aegis as a High Reliability Organization," *The Navalist*, September 16, 2017, https://thenavalist.com/home/2017/9/16/profound-simplicity-part-i-aegis-as-a-high-reliability-organization, 4; Friedman, "Missiles in Fleet Air Defense," 296; Emch, "Air Defense for the Fleet," 49; Robert E. Gray and Troy S. Kimmel, "The AEGIS Movement," *Naval Engineers Journal* 121, no. 3 (October 5, 2009): 39; Flanagan and Sweet, "Aegis: Advanced Surface Missile System," *Johns Hopkins APL Technical Digest* 2, no. 4 (1981): 244; Threston, "The Aegis Weapon System," 89; David L. Boslaugh, "First-Hand: Legacy of NTDS—Chapter 9 of the Story of the Naval Tactical Data System," *ETHW* (wiki), last modified May 12, 2021, https://ethw.org/First-Hand:Legacy_of_NTDS_-_Chapter_9_of_the_Story_of_the_Naval_Tactical_Data_System#Digitizing_Terrier; Meyer, *Reminiscences*, 375.
8. Emch, "Air Defense for the Fleet," 49; Chester C. Phillips, "Aegis: Advanced Multi-function Array Radar," *Johns Hopkins APL Technical Digest* 2, no. 4 (1981): 247.

CHAPTER 22. IMPLEMENTING THE AEGIS WEAPON SYSTEM

1. Meyer, *Reminiscences*, 303, 314; Morton, "Profound Simplicity, Part I."
2. Boslaugh, *When Computers Went to Sea*, 382; Meyer, "Our Navy—Like Our Lives—Is Continuous," 111; Boslaugh, *When Computers Went to Sea*, 197.
3. Gray and Kimmel, "The AEGIS Movement," 39–41; Wayne Meyer, as quoted in Gray and Kimmel, "The AEGIS Movement," 41.
4. John F. Lehman Jr., *Command of the Seas* (Annapolis, MD: Naval Institute Press, 2001), 155.
5. Gray and Kimmel, "The AEGIS Movement," 42; Terry C. Pierce, *Warfighting and Disruptive Technologies* (London: Frank Cass, 2004), 166; John L. Fialka, "New Antimissile Ship Faces Further Storms as Costs, Doubts Grow," *Wall Street Journal*, June 30, 1983, 1, 24; Thomas C. Hone, "The Program Manager as Entrepreneur: AEGIS and RADM Wayne Meyer," *Defense Analysis* 3, no. 3 (1987): 198.
6. Hone, "The Program Manager as Entrepreneur," 198–99; Muir, *Black Shoes and Blue Water*, 198; Friedman, "Missiles in Fleet Air Defense," 280.
7. "DD-963 *Spruance*-class," *Global Security*, accessed July 11, 2019, https://www.globalsecurity.org/military/systems/ship/dd-963-history.htm; Gray and Kimmel, "The AEGIS Movement," 43; U.S. General Accounting Office, *DLGN-38 Nuclear Guided Missile Frigate*, March 1974, 2–3, https://gao.justia.com/department-of-defense/1974/3/dlgn-38-nuclear-guided-missile-frigate-093821/093821-full-report.pdf.
8. Muir, *Black Shoes and Blue Water*, 198; Harlan Ullman, "Now Here This—We Need a New Project 60," U.S. Naval Institute *Proceedings*, 139, no. 12 (December 2013), https://www.usni.org/magazines/proceedings/2013-12/now-hear-we-need-new-project-60; Pierce, *Warfighting and Disruptive Technologies*, 159; Hone, "The Program Manager as Entrepreneur,"199.
9. James D. Flanagan and George W. Luke, "Aegis: Newest Line of Navy Defense," *Johns Hopkins APL Technical Digest* 2, no. 4 (1981): 239; Jerry E. Wacker, "A Managerial Approach to the Determination and Selection of the Tactical Combat System for Surface Navy Ships," DTIC ADA039674 (Defense Systems Management School, Port Belvoir, VA, May 1975), 15, https://archive.org/details/DTIC_ADA039674/page/n3/mode/2up; Muir, *Black Shoes and Blue Water*, 215; Testimony of Dr. Robert A. Forsch, Assistant Secretary of the Navy for R&D, in U.S. House of Representatives, *Hearings on Military Posture and H.R. 3818 and H.R. 8687 . . . before the Committee on Armed Services House of Representatives*, 92nd Cong., 1st Sess., Part 2 (Washington, DC: Government Printing Office, 1971), 4196.
10. Hone, "The Program Manager as Entrepreneur," 199.
11. Hone, "The Program Manager as Entrepreneur," 199; Meyer, *Reminiscences*, 444. Bagley was Zumwalt's principal assistant according to Muir, *Black Shoes and Blue Water*, 196.
12. Meyer, *Reminiscences*, 444.
13. Meyer, *Reminiscences*, 445.
14. Hone, "The Program Manager as Entrepreneur," 199; Klingaman, *APL—Fifty Years of Service to the Nation*, 194. For a detailed discussion of the shipbuilding and funding issues facing Admiral Zumwalt, which is beyond the scope of this book, see Muir, *Black Shoes and Blue Water*, 187–91, 197–198, 212.
15. Threston, "The AEGIS Weapon System," 92.

16. U.S. Senate, *Department of Defense Appropriations for Fiscal Year 1973, Hearings before a Subcommittee on Appropriations United States Senate . . . Part 3*, 96th Cong., 1st Sess., (Washington, DC: Government Printing Office, 1979), 1169–70; Flanagan and Luke, "Aegis: Newest Line of Navy Defense," 239; Norman Polmar, "The Typhon That Never Was," *Naval History* 31, no. 3 (June 2017): 10–11; Gray and Kimmel, "The AEGIS Movement," 43.
17. Gray and Kimmel, "The AEGIS Movement," 45; Hone, "The Program Manager as Entrepreneur," 200.
18. Todd Blades, "DDG-47: Aegis on Its Way to Sea," U.S. Naval Institute *Proceedings* 105, no. 1 (January 1985): 102; Morton, "Profound Simplicity, Part I," 5.
19. Boslaugh, *When Computers Went to Sea*, 126.
20. "Conference Report on H.R. 14592, Defense Procurement Appropriations," 93rd *Congressional Record–House*, July 24, 1974, 24942.
21. Thomas C. Hone, Douglas V. Smith, and Roger C. Easton Jr., "Aegis–Evolutionary or Revolutionary Technology?," in *The Politics of Naval Innovation*, edited by Bradd C. Hayes and Douglas Smith (Newport, RI: Center for Naval Warfare Studies, U.S. Naval War College, 1994), 47; U.S. Senate, Subcommittee of the Committee on Appropriations, *Hearings before a Committee of the Committee on Appropriations United States Senate Ninety-Third Congress, First Session: An Act Making Appropriations for the Department of Defense for the Fiscal Year Ending June 30, 1974, and for Other Purposes,* Part 3, Department of the Navy (Washington, DC: Government Printing Office, 1973), 1310; Muir, *Black Shoes and Blue Water*, 207, 212; Hone, "The Program Manager as Entrepreneur," 201; Boslaugh, *When Computers Went to Sea*, 123.
22. "Conference Report on H.R. 14592," 24936; Hone, Smith, and Easton, "Aegis–Evolutionary or Revolutionary Technology," 47.
23. Hone, Smith, and Easton, "Aegis–Evolutionary or Revolutionary Technology," 48.
24. Hone, "The Program Manager as Entrepreneur," 202.
25. Hone, "The Program Manager as Entrepreneur," 200; Meyer, *Reminiscences*, 383.
26. Threston, "The AEGIS Weapon System," 98–99.
27. Threston, "The AEGIS Weapon System," 98–99; Wray W. Bridger and Mark D. Ruiz, "Total Ownership Cost Reduction Study: Aegis Radar Phase Shifters" (Naval Postgraduate School, Monterey, CA, 2016), 18; Norman R. Landry, Hunter C. Goodrich, Henry F. Inacker, and Louis J. Lavedan Jr., "Aspects of Phase-Shifter and Driver Design for a Multifunction Phased-Array RADAR System," *IEEE Transactions on Microwave Theory and Techniques* 22, no. 6 (June 1974): 618, 622, https://doi.org/10.1109/TMTT.1974.1128303; Meyer, *Reminiscences*, 429–30. As Bridger and Ruiz have noted in "Total Ownership Cost Reduction": "Detailed engineering/production data of the AEGIS Weapon System transition from the EDM-1 in the AN/SPY-1A is virtually non-existent or not available" (18).

CHAPTER 23. GETTING AEGIS TO SEA

1. Gray and Kimmel, "The Aegis Movement, 46–47.
2. Hone, "The Program Manager as Entrepreneur," 203.
3. Randall H. Fortune and Brian T. Perkinson, "Getting Aegis to Sea: The Aegis Ships," *Naval Engineers Journal* 121, no. 3 (October 2009): 158.

4. Hone, Smith, and Easton, "Aegis—Evolutionary or Revolutionary Technology,"49; Blades, "DDG 47," 104.
5. James H. Doyle, "The Aegis Movement: An OPNAV Perspective," *Naval Engineers Journal* 121, no. 3 (October 2009): 198.
6. Doyle, "The Aegis Movement: An OPNAV Perspective," 197.
7. Meyer, *Reminiscences*, 334.
8. Doyle, "The Aegis Movement: An OPNAV Perspective," 199.
9. Gray and Kimmel, "The AEGIS Movement," 47; Doyle, "The Aegis Movement: An OPNAV Perspective," 200; Blades, "DDG 47," 104.
10. Hone, Smith, and Easton, "Aegis—Evolutionary or Revolutionary Technology," 51.
11. Hone, Smith, and Easton, "Aegis—Evolutionary or Revolutionary Technology," 52; see also Hone, "The Program Manager as Entrepreneur," 203.
12. Hone, Smith, and Easton, "Aegis—Evolutionary or Revolutionary Technology," 53.
13. Hone, Smith, and Easton, "Aegis—Evolutionary or Revolutionary Technology," 54.
14. Robert C. Staiman, "Aegis Cruiser Weight and Reduction Control," *Naval Engineers Journal* 99, no. 3 (May 1987): 190.
15. Hone, Smith, and Easton, "Aegis—Evolutionary or Revolutionary Technology," 1.
16. Blades, "DDG 47," 101; Gray and Kimmel, "The AEGIS Movement," 48; "Aegis DDG Contract Signed: 47 Coming On," *Surface Warfare* 4, no. 2 (February 1979): 10.
17. Robert C. Staiman, "Aegis Cruiser Weight Reduction and Control," 192.
18. "CG-47—USS *Ticonderoga*," *Seaforces*, accessed December 12, 2022, https://www.seaforces.org/usnships/cg/CG-47-USS-Ticonderoga.htm; Naval History and Heritage Command, "USS *Ticonderoga* (CG-47): 1983–2004," *DANFS online*, accessed December 12, 2022, https://www.history.navy.mil/research/histories/ship-histories/danfs/t/ticonderoga-v.html.

CHAPTER 24. THE TROUBLESOME *TICONDEROGA*

1. Gray and Kimmel, "The Aegis Movement," 48–49; Staiman, "Aegis Cruiser Weight Reduction and Control," 190–92; Edward L. Beach, letter to the editor, U.S. Naval Institute *Proceedings* 108, no. 10 (October 1985): 134; Richard Bard, "CG-47: Overweight and 'Ineffectual,'" *Defense Week*, August 16, 1982, as cited by Gray and Kimmel, "The Aegis Movement," 49.
2. Fialka, "New Antimissile Ship Faces Further Storms," 1, 24; Statement of Congressman Charles E. Bennert, March 1, 1983, U.S. House of Representatives, *Defense Department Authorization and Oversight Hearings on H.R. 2287 [H.R. 2969] Department of Defense Appropriations for Fiscal Year 1984 and Oversight of Previously Authorized Programs before the Committee on Armed Services House of Representatives*, 98th Cong., 1st Sess., Part 4, Seapower and Strategic and Critical Materials Subcommittee Title I (Washington, DC: Government Printing Office, 1983), 389; Vice Adm. R. L. Waters, "Overview of Shipbuilding Programs and Surface Ship Programs," in U.S. House of Representatives, *Defense Department Authorization and Oversight Hearings on H.R. 2287*, 97, 388.
3. Fialka, "New Antimissile Ship Faces Further Storms," 24.
4. Statement of the Assistant Secretary of the Navy (Research and Engineering and Systems), U.S. House of Representatives, *Department of Defense Appropriations for 1986, Hearings*

before a Subcommittee on Appropriations House of Representatives, 99th Cong., 1st Sess., Part 7 (Washington, DC: Government Printing Office, 1985), 963.

5. "*Ticonderoga* and the Aegis System," *NavSource Online: Battleship Photo Archive*, accessed December 12, 2022, http://www.navsource.org/archives/01/57s5.htm.
6. Threston, "The AEGIS Weapon System," 90.
7. Threston, "The AEGIS Weapon System," 90.
8. Threston, "The AEGIS Weapon System," 91–92.
9. Doyle, "The Aegis Movement: An OPNAV Perspective," 200.
10. Hone, Smith, and Easton, "Aegis—Evolutionary or Revolutionary Technology," 55–56.
11. Meyer, *Reminiscences*, 628, 630.
12. Meyer, *Reminiscences*, 363.
13. U.S. Navy, *Department of the Navy RDT&E Guide* (Washington, DC: Government Printing Office, 1979), E-29; Rear Admiral Wayne Eugene Meyer U.S. Navy (Ret.), Chronological Record of Service," in Meyer, *Reminiscences*, 2.

CHAPTER 25. REICH AND MEYER

1. "*Pluck*, *Pogy*, and *Portland*: Naming Navy Ships in World War II," *National World War II Museum*, accessed May 25, 2022, https://www.nationalww2museum.org/war/articles/naming-navy-ships-in-world-war-ii.
2. U.S. Navy, *A Report on Policies and Practices of the U.S. Navy for Naming the Vessels of the Navy* (Washington, DC: Department of the Navy, 2013), 8–9.
3. Joe Holley, "Wayne E. Meyer; Father of the Navy's Aegis Weapons System; 83," *San Diego Union-Tribune*, September 19, 2009, https://www.sandiegouniontribune.com/sdut-wayne-e-meyer-2009sep19-story.html; Dennis Mclellan, "Wayne E. Meyer Dies at 83; Retired Navy Rear Admiral," *Los Angeles Times*, September 2, 2009, https://www.latimes.com/local/obituaries/la-me-wayne-meyer2–2009sep02-story.html.

BIBLIOGRAPHY

BOOKS

Alden, John D. *The Fleet Submarine in the U.S. Navy: A Design and Construction History*. Annapolis, MD: Naval Institute Press, 1979.

Bash, Kevin, and Angelique Bash. *A Brief History of NORCO*. Charleston, SC: History Press, 2013.

Bauer, Theodore W. *History of the Industrial College of the Armed Forces*. Washington, DC: Alumni Association of the Industrial College of the Armed Forces, 1983.

Bijker, Wiebe E., Thomas Hughes, and T. J. Pinch, eds. *Social Construction of Technological Systems: New Directions in the Sociology and History of Technology*. Cambridge, MA: MIT Press, 1987.

Blair, Clay, Jr. *Silent Victory: The U.S. Submarine War against Japan*. New York: J. B. Lippincott, 1975.

Boslaugh, David L. *When Computers Went to Sea: The Digitization of the United States Navy*. Los Alamitos, CA: IEEE Computer Society, 1999.

Burgess, Eric. *Guided Weapons*. New York: Macmillan, 1957.

Fey, Peter. *Bloody Sixteen: The USS* Oriskany *and Air Wing 16 during the Vietnam War*. Lincoln, NE: Potomac Books 2018.

Fowler, Mary Lee. *Full Fathom Five: A Daughters' Search*. Tuscaloosa: University of Alabama Press, 2008.

Friedman, Norman. *Naval Radar*. Annapolis, MD: Naval Institute Press, 1981.

———. *U.S. Cruisers: An Illustrated Design History*. Annapolis, MD: Naval Institute Press, 1984.

———. *U.S. Destroyers: An Illustrated Design History*. Annapolis, MD: Naval Institute Press, 1982.

———. *U.S. Naval Weapons: Every Gun, Missile, Mine, and Torpedo Used by the U.S. Navy from 1883 to the Present Day*. Annapolis, MD: Naval Institute Press, 1988.

Green, Michael. *Images of War: United States Navy Destroyers Rare Photographs from Wartime Archives*. Bamsley, UK: Pen and Sword Maritime, 2020.

Hanser, Lawrence M., Louis W. Miller, Herbert J. Shukiar, and Bruce Newsome. *Developing Senior Navy Leaders: Requirements for Flag Officer Expertise Today and in the Future*. Santa Monica, CA: RAND Corporation, 2008.

Herge, Henry C. *Navy V-12*. Paducah, KY: Turner Publishing, 1996.

Hornfischer, James D. *Ship of Ghosts: The Story of the USS* Houston, *FDR's Legendary Lost Cruiser, and the Epic Saga of Her Survivors*. New York: Bantam, 2007.

Hughes, Thomas P. *Rescuing Prometheus*. New York: Pantheon Books, 1998.

Johns Hopkins University Applied Physics Laboratory. *The First Forty Years; A Pictorial Account of the Johns Hopkins University Applied Physics Laboratory*. Baltimore, MD: Johns Hopkins University Applied Physics Laboratory, 1983.

Klingaman, William K. *APL—Fifty Years of Service to the Nation: A History of the John Hopkins University Applied Physics Laboratory*. Laurel, MD: Johns Hopkins University Applied Physics

Laboratory, 1993. https://archive.org/details/aplfiftyyearsofs0000klin/page/n3/mode/2up.

Lehman, John F., Jr. *Command of the Seas*. Annapolis, MD: Naval Institute Press, 2001.

McBurney, Christian, Brian L. Wallin, Patrick T. Conley, John W. Kennedy, and Maureen A. Taylor. *World War II Rhode Island*. Charleston, SC: History Press, 2017. https://www.google.com/books/edition/World_War_II_Rhode_Island/iELFDgAAQBAJ?hl=en&gbpv=1&pg=PT6&printsec=frontcover.

Merrill, John, and Lionel D. Wyld. *Meeting the Submarine Challenge: A Short History of the Naval Underwater Systems Center*. Washington, DC: Government Printing Office, 1997. https://www.google.com/books/edition/Meeting_the_Submarine_Challenge/tO53-JeqP5EC?hl=en.

Michno, Gregory F. *Death on the Hellships: Prisoners at Sea in the Pacific War*. Annapolis, MD: Naval Institute Press, 2001.

——. *USS* Pampanito *Killer-Angel*. Norman: University of Oklahoma Press, 2001.

Mindell, David A. *Between Human and Machine: Feedback, Control, and Computing before Cybernetics*. Baltimore, MD: Johns Hopkins University Press, 2002.

Muir, Malcolm, Jr., *Black Shoes and Blue Water: Surface Warfare in the United States Navy, 1945–1975*. Washington, DC: Naval Historical Center, 1996.

Ordway, Frederick I., III, and Ronald C. Wakeford. *International Missile and Spacecraft Guide*. New York: McGraw-Hill, 1960.

Pierce, Terry C. *Warfighting and Disruptive Technologies*. London: Frank Cass, 2004.

Polmar, Norman, and K. J. Moore. *Cold War Submarines: The Design and Construction of U.S. and Soviet Submarines*. Dulles, VA: Potomac Books, 2004.

Polmar, Norman, and Robert S. Norris. *The U.S. Nuclear Arsenal: A History of Weapons and Delivery Systems since 1945*. Annapolis, MD: Naval Institute Press, 2009.

Rubel, John H. *Memoirs III: Time and Chance 1959–1976*. Santa Fe, NM: Kay Say Publications, 2006.

Slason, Frank. K., ed. *The Lucky Bag*. Annapolis, MD: Class of 1935, 1935.

Smith, Hayes, ed. *The Politics of Naval Innovation*. Newport, RI: Center for Naval Warfare Studies, U.S. Naval War College, 1994.

Spinaldi, Graham. *From Polaris to Trident: The Development of US Fleet Ballistic Missile Technology*. Cambridge, UK: Cambridge University Press, 1994.

Weir, Gary E. *Building American Submarines 1914–1940*. Washington, DC: Naval Historical Center, 1991.

——. *Forged in War: The Naval-Industrial Complex and American Submarine Construction, 1940–1961*. Washington, DC: Naval Historical Center, 1993.

Werrell, Kenneth P. *Archie to SAM: A Short Operational History of Ground-Based Air Defense*. Maxwell Air Force Base, AL: Air University Press, 2005.

Wildenberg, Thomas. *All the Factors of Victory: Joseph Mason Reeves and the Origins of Carrier Air Power*. Annapolis, MD: Naval Institute Press, 2018.

Wildenberg, Thomas, and Norman Polmar. *Ship Killers: A History of the American Torpedo*. Annapolis, MD: Naval Institute Press, 2010.

Willis, Ron L., and Thomas Carmichael. *United States Navy Wings of Gold From 1917 to the Present*. Atglen, PA: Schiffer, 1995.

Wolfe, Audra J. *Competing with the Soviets: Science, Technology and the State in Cold War America*. Baltimore, MD: Johns Hopkins University Press, 2013.

ARTICLES

"3"/50 (7.62 cm) Marks 27, 33 and 34." *NavWeaps*, accessed September 4, 2021. http://www.navweaps.com/Weapons/WNUS_3-50_mk27-33-34.php.

"ADM James Fife, Jr. (1897–1975)." *Find a Grave Memorial*, accessed July 14, 2022. https://www.findagrave.com/memorial/6800842/james-fife.

"Aegis DDG Contract Signed: 47 Coming On." *Surface Warfare* 4, no. 2 (February 1979): 10–11.

"Albert Kingsbury." *National Inventors Hall of Fame*, accessed July 13, 2022. https://www.invent.org/inductees/albert-kingsbury.

Alison, Carolyn. "V-12: the Navy College Training Program." *RootsWeb*, accessed May 28, 2022. https://homepages.rootsweb.com/~uscnrotc/V-12/.

"AN/SPS-4." *Radartutorial*, accessed September 12, 2022. https://www.radartutorial.eu/19.kartei/11.ancient2/karte025.en.html.

Bard, Richard. "CG-47: Overweight and 'Ineffectual.'" *Defense Week*, August 16, 1982, 1.

Berl, Walter G. "Annotated Bumblebee Initial Report." *Johns Hopkins APL Technical Digest* 3, no. 2 (1982): 171–79. https://secwww.jhuapl.edu/techdigest/Content/techdigest/pdf/V03-N02/03-02-Berl.pdf.

Berry, William W. "Radars and Computers aboard USS *Sterett* and Other Warships during the Vietnam War." *NavWeaps*, accessed September 30, 2022. http://www.navweaps.com/index_tech/tech-088.php.

Blades, Todd. "DDG-47: Aegis on Its Way to Sea." U.S. Naval Institute *Proceedings* 105, no. 1 (January 1985): 101–5.

Boslaugh, David L. "First-Hand: Legacy of NTDS—Chapter 9 of the Story of the Naval Tactical Data System." *ETHW* (wiki), last modified May 12, 2021. https://ethw.org/First-Hand:Legacy_of_NTDS_-_Chapter_9_of_the_Story_of_the_Naval_Tactical_Data_System#Digitizing_Terrier.

———. "First-Hand: Moving the Firing Key to NTDS—Chapter 6 of the Story of the Naval Tactical Data System." *ETHW* (wiki), last modified May 12, 2021. https://ethw.org/First-Hand:Moving_the_Firing_Key_to_NTDS_-_Chapter_6_of_the_Story_of_the_Naval_Tactical_Data_System.

———. "First-Hand: No Damned Computer Is Going to Tell Me What to DO—The Story of the Naval Tactical Data System, NTDS." *ETHW* (wiki), last modified May 12, 2021. http://ethw.org/First-Hand:No_Damned_Computer_is_Going_to_Tell_Me_What_to_DO_-_The_Story_of_the_Naval_Tactical_Data_System,_NTDS.

"CG-47—USS *Ticonderoga*." *Seaforces*, accessed December 12, 2022. https://www.seaforces.org/usnships/cg/CG-47-USS-Ticonderoga.htm.

Chant, Chris. "First of the Great Naval SAMs—the RIM-2 Terrier." *Chris Chant's Blog*, March 10, 2011. https://cmchant.com/the-rim-2-terrier-the-first-of-the-great-naval-sams/

"Charles Andrews Lockwood (1890–1967)." *One-Name Net*, accessed April 4, 2023. https://lockwood.one-name.net/Charles%20A%20Lockwood.htm.

Cockell, William A., Jr. "The DLG Modernization Program." U.S. Naval Institute *Proceedings* 98, no. 1 (January 1972): 101–4.

Collins Radio Co. *HF Communication Equipment* [catalog] *1984–85*.

Cooper, Kenneth G. "Naval Ship Production: A Claim Settled and a Framework Built." *Interfaces* 10, no. 6 (December 1980): 22.

D'Amico, Angela, and Richard Pittenger. "A Brief History of Active Sonar." *Aquatic Mammals* 35, no. 4 (2009): 426–34. https://csi-test.whoi.edu/sites/default/files/literature/Full%20Text/index.pdf.

Dean, Frank. "The Unified Talos." *Johns Hopkins APL Technical Digest* 3, no. 2 (1982): 123–24.

"DD-963 *Spruance*-class." *Global Security*, accessed July 11, 2019. https://www.globalsecurity.org/military/systems/ship/dd-963-history.htm.

Dienesch, Robert. "Radar and the American Submarine War, 1941–1945: A Reinterpretation." *Northern Mariner/Le marin du nord* 14, no. 3 (July 2004): 27–40.

Domalgalski, John J. "Disaster at Cavite." *Naval History Magazine* 32, no. 6 (December 2018). https://www.usni.org/magazines/naval-history-magazine/2018/december/disaster-cavite.

Doyle, James H., Jr. "The Aegis Movement: An OPNAV Perspective." *Naval Engineers Journal* 121, no. 3 (October 2009): 197–201.

Emch, George F. "Air Defense for the Fleet." *Johns Hopkins APL Technical Digest* 13, no. 1 (1992): 39–56.

Feldhaus, Jack. "Rescue of LCDR Thomas A. Tucker 31 August 1966 Haiphong Harbor." CDR Jack Feldhaus, USN, KIA 8 October 1966—Memorial Web Site, accessed October 20, 2022. https://1feldhaus.tripod.com/cmdrjackfeldhaususn/id47.html.

Ferrara, Joe. "DoD's 5000 Documents: Evolution and Change in Defense Acquisition Policy." *Acquisition Review Quarterly* (Fall 1996): 109–30. https://www.dau.edu/library/arj/ARJ/arq96/ferrar.pdf.

Fialka, John L. "New Antimissile Ship Faces Further Storms as Costs, Doubts Grow." *Wall Street Journal*, June 30, 1983, p. 1, 24.

Flanagan, James D., and George W. Luke. "Aegis: Newest Line of Navy Defense." *Johns Hopkins APL Technical Digest* 2, no. 4 (1981): 237–42.

Flanagan, James D., and William N. Sweet. "Aegis: Advanced Surface Missile System." *Johns Hopkins APL Technical Digest* 2, no. 4 (1981): 243–45.

"Fort Bliss Research and Training." *TheMilitaryStandard*, accessed August 28, 2022. http://www.themilitarystandard.com/missile/fortbliss.php.

Fortune, Randall H., and Brian T. Perkinson. "Getting Aegis to Sea: The Aegis Ships." *Naval Engineers Journal* 121, no. 3 (October 2009): 155–75.

Fredrickson, John, and John Roper. "The Kansas City B-25 Factory." *Air and Space Magazine*, June 11, 2014. https://www.smithsonianmag.com/air-space-magazine/kansas-city-b-25-factory-180951624/.

Garten, William, Jr., and Frank A. Dean. "Evolution of the Talos Missile." *Johns Hopkins APL Technical Digest* 3, no. 2 (1982): 117–22.

Gibson, Ralph E. "Alexander Kossiakoff: His Life and Career, Part One." *Johns Hopkins APL Technical Digest* 27, no. 1 (2006): 1–11.

Grant, Lachlan. "70th Anniversary of the Sinking of the *Rakuyo Maru*." *Australian War Memorial*, accessed August 8, 2022. https://www.awm.gov.au/articles/blog/70th-anniversary-sinking-rakuy-maru.

Gray, Robert E., and Troy S. Kimmel. "The AEGIS Movement." *Naval Engineers Journal* 121. no. 3 (October 5, 2009): 37–69.

Gregory, Mackenzie J. "Top Ten US Submarine Captains in WW2 By Number of Ships Sunk." *Ahoy—Mac's Web Log*, accessed August 4, 2022. http://www.ahoy.tk-jk.net/macslog/TopTenUSNavySubmarineCapt.html.

Gropman, Alan L. "Industrial College of the Armed Forces: A Primer." *National Defense*, January 1, 2008. https://www.nationaldefensemagazine.org/articles/2007/12/31/2008january-industrial-college-of-the-armed-forces-a-primer#.

Gulick, Joseph W., Coleman Hyatt, and Oscar M. Martin Jr. "The Talos Guidance System." *Johns Hopkins APL Technical Digest* 3, no. 2 (1982): 243–53.

Gussow, Milton, and Edward C. Prettyman. "Typhon: A Weapon System ahead of Its Time." *Johns Hopkins APL Technical Digest* 13, no. 1 (1992): 143–89.

Hackett, Bob. "*Harbin Maru*-Class Auxiliary Hospital Ship/Transport." *Combined Fleet*, accessed July 17, 2022. http://www.combinedfleet.com/Harbin_c.htm.

"The Hall of Valor Project." *Military Times*, accessed August 4, 2022. https://valor.militarytimes.com/hero/20804.

Harpster, Joseph W. C. "USS *Stoddard* DD566 in the Korean War Era." *USS Stoddard*, accessed September 3, 2022. http://www.ussstoddard.org/koreaharpster.html.

Hawes, Ralph E., and James R. Whalen. "The Standard Missile/AEGIS Story." *Naval Engineers Journal* 121, no. 3 (October 5, 2009): 133–54.

Hays, Phillip R. "Talos Missile History." *Okieboat*, accessed September 9, 2022. https://www.okieboat.com/Talos%20history.html.

———. "USS *Oklahoma City* CL91/CLG5/CG5 SPG-49 Talos Mk 77 Guided Missile Fire Control System." *Okieboat*, accessed December 3, 2022. https://www.okieboat.com/Talos%20fire%20control%20system.html.

———. "USS *Oklahoma City* CL91/CLG5/CG5 SPG-49 Talos Mk 77 Guided Missile Launching System." *Okieboat*, accessed September 9, 2022. https://www.okieboat.com/Talos%20launching%20system.html.

———. "USS *Oklahoma City* CL91/CLG5/CG5 SPG-49 Tracking Radar." *Okieboat*, accessed September 9, 2022. https://www.okieboat.com/SPG-49%20description.html.

Holley, Joe. "Wayne E. Meyer; Father of the Navy's Aegis Weapons System; 83." *San Diego Union-Tribune*, September 19, 2009. https://www.sandiegouniontribune.com/sdut-wayne-e-meyer-2009sep19-story.html.

Hone, Thomas C. "The Program Manager as Entrepreneur: AEGIS and RADM Wayne Meyer." *Defense Analysis* 3, no. 3 (1987): 197–212.

Hone, Thomas C., Douglas V. Smith, and Roger C. Easton Jr. "Aegis—Evolutionary or Revolutionary Technology?" In *The Politics of Naval Innovation*, edited by Bradd C. Hayes and Douglas Smith, 38–66. Newport RI: Center for Naval Warfare Studies, U.S. Naval War College, 1994.

"IJN Minelayer *Shirataka*: Tabular Record of Movement." *Combinedfleet*, accessed August 1, 2022. http://www.combinedfleet.com/Shirataka_t.htm.

Inman, Bruce D. "From Typhon to Aegis—The Issues and Their Resolution." *Naval Engineers Journal* 100, no. 3 (May 1988): 62–72.

Irvine, Robert K. "Surface Missile Systems Management Action Keeps Pace with Accelerating Technology." *Navy Management Review* 12, no. 8 (August 1967): 28–31.

"James Fife." *Hall of Valor Project*, accessed July 14, 2022. https://valor.militarytimes.com/hero/27313.

"John T. Hayward." *DBpedia*, accessed September 9, 2022. https://dbpedia.org/page/John_T._Hayward.

Johnston, David L. "A Visual Guide to the S-Class Submarines 1918–1945, Part 1: The Prototypes." *Submarine Museums*, accessed August 2, 2023. https://www.submarinemuseums.org/docs/sboats1_v2.pdf.

"*Kachidoki Maru*." *POW Research Network Japan*, accessed August 8, 2022. http://www.powresearch.jp/en/archive/ship/kachidoki.html.

Krige, John. "The 1984 Physics Prize for Heterogeneous Engineering." *Minerva* 39, no. 4 (December 2001): 425–43.

Landry, Norman R., Hunter C. Goodrich, Henry F. Inacker, and Louis J. Lavedan. "Practical Aspects of Phase-Shifter and Driver Design for a Multifunction Phased-Array RADAR System." *IEEE Transactions on Microwave Theory and Techniques* 22, no. 6 (June 1974): 617–25. https://doi.org/10.1109/TMTT.1974.1128303.

Lentinello, Richard. "How Gas Rationing Worked during World War II." *Hemings*, accessed July 24, 2022. https://www.hemmings.com/stories/2019/12/30/how-gas-rationing-worked-during-world-war-ii.

Lockheed Martin. "Aegis Heritage," presentation, November 20, 2002. As cited in "Aegis Combat System." *Wikipedia*, accessed November 13, 2022. https://en.wikipedia.org/wiki/Aegis_Combat_System.

"Lockwood, Charles Andrew, Jr. (1890–1967)." *The Pacific War Online Encyclopedia*, by Kent G. Budge, 2016. http://pwencycl.kgbudge.com/L/o/Lockwood_Charles_A.htm.

Mackaman, Tom. "From Greece to Ukraine: 75 Years of the Truman Doctrine." *World Socialist Web Site*, June 15, 2022. https://www.wsws.org/en/articles/2022/06/16/uyno-j16.html.

Marchio, James. "The Evolution and Relevance of Joint Intelligence Centers." *Studies In Intelligence* 48, no. 1 (2004): 41–51.

Mclellan, Dennis. "Wayne E. Meyer Dies at 83; Retired Navy Rear Admiral." *Los Angeles Times*, September 2, 2009. https://www.latimes.com/local/obituaries/la-me-wayne-meyer2–2009sep02-story.html.

Meyer, Wayne E. "Our Navy—Like Our Lives—Is Continuous." *Johns Hopkins APL Technical Digest* 23, nos. 2– 3 (2002): 111–13.

Milford, Frederick. "U.S. Navy Torpedoes Part Five: Post WW-II Submarine Launched Heavyweight Torpedoes." *Submarine Review*, October 1997, 70–86.

Montoya, Matthew. "Standard Missile: A Cornerstone of Navy Theater Air Missile Defense." *Johns Hopkins APL Technical Digest* 22, no. 3 (2001): 234–47.

Morton, John F. "Profound Simplicity, Part I: Aegis as a High Reliability Organization." *The Navalist*, September 16, 2017. https://thenavalist.com/home/2017/9/16/profound-simplicity-part-i-aegis-as-a-high-reliability-organization.

"National Security Industrial Association." *NDIA Rocky Mountain*, accessed September 11, 2022. https://ndiarmc.org/about/history-of-ndia/.

"Naval Air Station Oakland." *Historic California Posts, Camps, Stations and Airfields*, accessed August 12, 2022. http://www.militarymuseum.org/NASOakland.html.

"Naval Surface Warfare Center (NSWC) Corona Division." *GlobalSecurity*, accessed October 8, 2022. https://www.globalsecurity.org/military/facility/corona.htm.

Oliver, Marion E. "Terrier/Tartar: Pacing the Threat." *Johns Hopkins APL Technical Digest* 2, no. 4 (1981): 256–60.

Oliver, Marion E., and William N. Sweet. "Standard Missile: The Common Denominator." *Johns Hopkins APL Technical Digest* 2, no. 4 (1981): 284.

Paddison, Fletcher C. "The Talos Control System." *Johns Hopkins APL Technical Digest* 3, no. 2 (1982): 154–56.

Parsch, Andreas. "RIM-2." *Directory of U.S. Military Rockets and Missiles*, 2004. http://www.designation-systems.net/dusrm/m-2.html.

——. "RIM-8." *Directory of U.S. Military Rockets and Missiles*, 2004. https://www.designation-systems.net/dusrm/m-8.html.

——. "SAM-N-2/SAM-N-2 [Lark]." *Directory of U.S. Military Rockets and Missiles*, 2004. https://www.designation-systems.net/dusrm/appl/sam-n-2.html.

Pearce, William. "Man Double-Acting Diesel Marine Engines." *Old Machine Press*, December 20, 2017. https://oldmachinepress.com/2017/12/20/man-double-acting-diesel-marine-engines/.

Phillips, Chester C. "Aegis: Advanced Multi-function Array Radar." *Johns Hopkins APL Technical Digest* 2, no. 4 (1981): 246–49.

"*Pluck*, *Pogy*, and *Portland*: Naming Navy Ships in World War II." *National WWII Museum*, accessed May 25, 2022. https://www.nationalww2museum.org/war/articles/naming-navy-ships-in-world-war-ii.

Polmar, Norman. "The Typhon That Never Was." *Naval History* 31, no. 3 (June 2017): 10–11.

"RADM Frederick Burdett Warder." *Military Hall of Honor*, accessed September 20, 2022. https://militaryhallofhonor.com/honoree-record.php?id=3195.

Robinson, Elmer D. "The Talos Ship System." *Johns Hopkins APL Technical Digest* 3, no. 2 (1982): 162–66.

Rogoway, Tyler. "The Talos Missile Had a Wonderfully Complex Shipboard Assembly Line of a Launch System." *The War Zone*, August 2, 2008. https://www.thedrive.com/the-war-zone/22599/the-talos-missile-had-a-wonderfully-complex-shipboard-assembly-line-of-a-launch-system.

Rutz, Paul X. "Crusader Down." *Historynet*, accessed November 19, 2022. https://www.historynet.com/crusader-down/.

"*Saikyo Maru* (1936–) *Saikyo Maru* (+ 1942)." *Wrecksite*, accessed July 7, 2022. https://www.wrecksite.eu/wreck.aspx?306650.

"SC Family of Radars." *Radartutorial*, accessed August, 26, 2022. https://www.radartutorial.eu/19.kartei/11.ancient/karte011.en.html.

Shaw, James C. "First Year of the Line School." U.S. Naval Institute *Proceedings* 73, no. 12 (November 1947): 1341–45.

Simons, William E., ed. *Professional Military Education in the United States: A Historical Dictionary*. Westport, CT: Greenwood Press, 2000.

"The Sinking of Prisoner of War Transport Ships in the Far East." *Imperial War Museums*, accessed August 3, 2022. https://www.iwm.org.uk/history/the-sinking-of.

Sires, Hunter. "They Were Playing Chicken: The U.S. Asiatic Fleet's Gray-Zone Deterrence Campaign against Japan, 1937–40." *Naval War College Review* 72, no. 3 (Summer 2019): 139–58.

Smolinski, Mike. "K-1 Barracuda (SSK-1) (SST-3)." *NavSource Online: Submarine Photo Archive*, accessed August 30, 2022. http://www.navsource.org/archives/08/08550.htm.

Staiman, Robert C. "Aegis Cruiser Weight Reduction and Control." *Naval Engineers Journal* 99, no. 3 (May 1987): 190–201.

"*Stingray* (SS 186)." *Uboat,* accessed July 10, 2022. https://uboat.net/allies/warships/ship/2938.html.

"Stuffing Box in Marine Diesel Engine." *The Marine Whales,* accessed June 6, 2022. https://themarinewhales.in/stuffing-box-in-marine-diesel-engine/.

Sundin, Sara. "Make It Do—Gasoline Rationing in World War II." *Sarah Sundin,* July 22, 2022. https://www.sarahsundin.com/make-it-do-gasoline-rationing-in-world-war-ii-2/.

Threston, Joseph T. "The AEGIS Weapon System." *Naval Engineers Journal* 121, no. 3 (October 5, 2009): 85–108.

"*Ticonderoga* and the Aegis System." *NavSource Online: Battleship Photo Archive,* accessed December 12, 2022. http://www.navsource.org/archives/01/57s5.htm.

"Truman Doctrine (1947)." Milestone Documents, *National Archives,* last updated February 8, 2022. https://www.archives.gov/milestone-documents/truman-doctrine.

"TRW Systems, Operating Group of TRW Inc." *Signal,* September 1966, p. 83.

Tully, Anthony P. "The Loss of Battleship KONGO: As Told in Chapter 'November Woes' of 'Total Eclipse: The Last Battles of the IJN—Leyte to Kure 1944 to 1945.'" *Combined-Fleet,* 1998. http://www.combinedfleet.com/eclipkong.html.

"Typhon DLG/DLGN/CGN/." *Global Security,* accessed October 5, 2022. https://www.globalsecurity.org/military/systems/ship/cgn-typhon.htm.

Ullman, Harlan. "Now Here This—We Need a New Project 60." U.S. Naval Institute *Proceedings* 139, no. 12 (December 2013). https://www.usni.org/magazines/proceedings/2013-12/now-hear-we-need-new-project-60.

"United States Asiatic Fleet Locations December 7, 1941." *NavSource,* accessed July 1, 2022. http://www.navsource.org/Naval/usfb.htm.

"U.S. Navy Hunter-Killer Submarines." *Weapons and Warfare,* accessed August 30, 2022. https://weaponsandwarfare.com/2017/01/25/us-navy-hunter-killer-submarines/.

Wallin, Brian L. "The Torpedo Station at Newport." In Christian McBurney et al. *World War II Rhode Island,* 19–30. Charleston, SC: History Press, 2017.

"Warder, Frederick Burdett, Radm." *Together We Served,* accessed November 21, 2022. https://navy.togetherweserved.com/usn/servlet/tws.webapp.WebApp?cmd=ShadowBoxProfile&type=EventExt&ID=229112.

"Wayne E. Meyer." *Chariton County Historical Society,* accessed July 23, 2022. https://m.facebook.com/charitoncountymuseum/photos/wayne-e-meyerwayne-e-meyer-was-born-in-brunswick-missouri-on-april-21-1926-to-eu/1792752124316360/.

"Why They Wear the Dolphins." *All Hands,* no. 608 (September 1967): 14–16.

DISSERTATIONS, PAPERS, REPORTS, AND UNPUBLISHED MATERIAL

Bridger, Wray W., and Mark D. Ruiz. "Total Ownership Cost Reduction Case Study: Aegis Radar Phase Shifters." Naval Postgraduate School, Monterey, CA, 2016.

Center for Naval Analysis. *Defense against Kamikaze Attacks in World War II and Relevance to Anti-Ship Missile Defense.* Vol. 1, *An Analytical History of Kamikaze Attacks against Ships of the United States Navy during World War II.* Alexandria, VA: Center for Naval Analysis, 1971. https://apps.dtic.mil/sti/citations/AD0725163.

Fahrney, Delmar S. "History of Pilotless Aircraft and Guided Missiles." Unpublished manuscript produced for the U.S. Navy. RG 72, National Archives, College Park, MD.

Friedman, Norman. "Missiles in Fleet Air Defense." Unpublished history prepared for the U.S. Navy. RG 72, National Archives, College Park, MD.

McGraw, Matthew Robert. "Beneath the Surface: American Culture and Submarine Warfare in the Twentieth Century." Master's thesis, University of Southern Mississippi, 2011. https://aquila.usm.edu/cgi/viewcontent.cgi?article=1233&context=masters_theses.

Manuel, Walter. "Who Becomes a Limited Duty Officer and Chief Warrant Officer? An Examination of Differences of Limited Duty Officers and Chief Warrant Officers." Thesis, U.S. Naval Postgraduate School, Monterey, CA, June 2006.

Nace, Wilbert J. "Term Project Report Systems Acquisition Management." Thesis, U.S. Naval Post Graduate School, Monterey, CA, January 1972.

Reich, Eli Thomas. "Transcript of Naval Service." National Military Records Center, National Archives, St. Louis, MO.

"USS *Kearsarge* (CVS 33) WestPac Cruise Book 1966." *Navy* [Federal Republic of Germany], accessed October 19, 2022. https://www.navysite.de/cruisebooks/cv33-66/index.html.

OFFICIAL REPORTS, GOVERNMENT PUBLICATIONS, AND PUBLIC DOCUMENTS

Annual Register of the United States Naval Academy Annapolis, M.D. September 29, 1933. Washington, DC.: Government Printing Office, 1934.

Annual Register of the United States Naval Academy Annapolis, M.D. September 28, 1934. Washington, DC: Government Printing Office, 1934.

Annual Register of the United States Naval Academy Annapolis, M.D. September 27, 1935. Washington, DC: Government Printing Office, 1936.

Armed Forces Staff College. *Command History of the Armed Forces Staff College 1946–1981*. Washington DC: Armed Forces Staff College, 1983[?].

Condit, Kenneth W. *History of the Joint Chiefs of Staff: The Joint Chiefs of Staff and National Policy*, vol. 2: *1947–1949*. Washington, DC: Office of Joint History, Office of the Chairman of the Joint Chiefs of Staff, 1996.

"Conference Report on H.R. 14592, Defense Procurement Appropriations." 93rd *Congressional Record–House*, July 24, 1975.

"Depth Charge Direction Indicator (DCDI)." Excerpt from *Catalogue of Naval Electronics Equipment-April 1946-NavShips 900,116*. San Francisco Maritime National Park Association, accessed July 6, 2023. https://maritime.org/tech/radiocat/dcdi.htm.

Furer, Julius. *Administrative History of Navy Department in World War II*. Washington, DC: Department of the Navy, 1959.

Gressett, Jessica, ed. *Army Air Artillery Defense School*. Fort Bliss, TX: U.S. Army, n.d.. https://srmsc.org/pdf/005240p0.pdf.

Hayes, Bradd C., and Douglas Smith, eds. *The Politics of Naval Innovation*. Newport, RI: Center for Naval Warfare Studies, U.S. Naval War College, 1994.

Joint Army-Navy Assessment Committee. *Naval and Merchant Shipping Losses during World War II by All Causes*. Washington, DC: Government Printing Office, 1947.

Jolie, E. W. *A Brief History of U.S. Navy Torpedo Development*. Newport, RI: Naval Underwater Systems Center, 1978.

Kanagy, Mary, ed. "BuOrd Relays CNO Praise for NOL Part in Special Job." *Report . . . U.S. Naval Ordnance Laboratory* 11, no. 10 (April 1955): 1.

"Louis Emil Denfeld (13 April 1891–28 March 1972)." *Arlington National Cemetery*, accessed August 11, 2022. https://www.arlingtoncemetery.net/ledenfeld.htm.

Manke, Robert C. "Overview of U.S. Navy Antisubmarine Warfare (ASW) Organization during the Cold War Era." NUWC-NPT Technical Report 11,890. Newport, RI: Naval Undersea Warfare Center Division, 2008.

Meyer, Wayne E. "Statement before U.S. Senate Subcommittee on Research and Development of the Committee on Armed Services." 96th U.S. Congress–Senate, April 3, 1979.

National Defense Research Committee. *Acoustic Torpedoes: Summary Technical Report of Division 6, NDRC*. Washington, DC: Office of Scientific Research and Development, 1946.

National Naval Aviation Museum. "AJ-2 Savage." *Naval History and Heritage Command*, accessed August 8, 2022. https://www.history.navy.mil/content/history/museums/nnam/explore/collections/aircraft/a/aj-2-savage0.html.

National Park Service. "USS *Pampanito*," accessed August 8, 2022. https://www.nps.gov/places/uss-pampanito.htm.

Naval Historical Center. *Dictionary of American Naval Fighting Ships* (*DANFS*), vol. 2. Washington, DC: Government Printing Office, 1963.

Naval Historical Center. *Dictionary of American Naval Fighting Ships* (*DANFS*), vol. 3. Washington, DC: Government Printing Office, 1968.

Naval Historical Center. *Dictionary of American Naval Fighting Ships* (*DANFS*), vol. 4. Washington, DC: Government Printing Office, 1969.

Naval Historical Center. *Dictionary of American Naval Fighting Ships* (*DANFS*), vol. 5. Washington, DC.: Government Printing Office, 1970.

Naval Historical Center. *Dictionary of American Naval Fighting Ships* (*DANFS*), vol. 6. Washington, DC: Government Printing Office, 1976.

Naval Historical Center. *Dictionary of American Naval Fighting Ships* (*DANFS*), vol. 7. Washington, DC: Government Printing Office, 1981.

Naval History and Heritage Command (NHHC). "*Aucilla* (AO-56): 1943–1976." *Dictionary of American Naval Fighting Ships* (*DANFS online*), accessed September 21, 2022. https://www.history.navy.mil/research/histories/ship-histories/danfs/a/aucilla.html.

Naval History and Heritage Command (NHHC). "*Compton* (DD-705)." *Dictionary of American Naval Fighting Ships* (*DANFS online*), accessed July 15, 2020. https://www.history.navy.mil/research/histories/ship-histories/danfs/c/compton.html.

Naval History and Heritage Command (NHHC). "*Stingray* (SS-186)." *Dictionary of American Naval Fighting Ships* (*DANFS online*), accessed November 17, 2022. https://www.history.navy.mil/content/history/nhhc/research/histories/ship-histories/danfs/s/stingray-ii.html.

Naval History and Heritage Command (NHHC). "USS *Ticonderoga* (CG-47): 1983–2004." *Dictionary of American Naval Fighting Ships* (*DANFS online*), accessed December 12, 2022. https://www.history.navy.mil/research/histories/ship-histories/danfs/t/ticonderoga-v.html.

Naval History and Heritage Command (NHHC). "*Tringa*." *Dictionary of American Naval Fighting Ships* (*DANFS online*), accessed September 21, 2022. https://www.history.navy.mil/research/histories/ship-histories/danfs/t/tringa.html.

Okula, Christopher. "Navy Celebrates 50 Years of Truth, Commemorates Golden Anniversary of Independent Assessment." *U.S. Navy*, accessed December 2, 2022. https://www.navy.mil/submit/display.asp?story_id=80607.

Poole, Walter S. *Adapting to Flexible Response 1960–1968: History of Acquisition in the Department of Defense, Volume II*. Washington, DC: Historical Office, Office of the Secretary of Defense, 2013.

Pyle, Russ. "Countdown in *Camden*." *Surface Warfare Magazine* 3, no. 2 (February 1978): 16–19.

Reich, E. T. "USS *Compton* (DD705) Factual History," September 1945.

Rife, James P., and Rodney P. Carlisle. *The Sound of Freedom: Naval Weapons Technology at Dahlgren, Virginia 1918–2006*. Dahlgren, VA: Naval Surface Warfare Center, 2006.

"Some Notes on the SP." *BuShips Electron* 1, no. 3 (September 1945): 16–19.

"Tested under Combat Conditions." *All Hands*, no. 601, February 1967, 24–30.

U.S. Department of State. *Sixth Report to Congress on Assistance to Greece and Turkey for the Period Ended December 31, 1948*. Washington, DC: Government Printing Office, 1949.

U.S. Fleet, Headquarters of the Commander in Chief. *Radar Bulletin No.3: Radar Operator's Manual*. Washington, DC: Navy Department, April 1945. https://www.ibiblio.org/hyperwar/USN/ref/RADTHREE/index.html.

U.S. General Accounting Office. *DLGN-38 Nuclear Guided Missile Frigate*. March 1974. https://gao.justia.com/department-of-defense/1974/3/dlgn-38-nuclear-guided-missile-frigate-093821/093821-full-report.pdf.

U.S. House of Representatives. *Department of Defense Appropriations for 1978 Hearings before a Subcommittee of the Committee on Appropriations House of Representatives*. 95th Cong., 1st Sess., Ingalls Shipyard/Litton Industries. Washington, DC: Government Printing Office, 1977.

U.S. House of Representatives. *Department of Defense Appropriations for 1986: Hearings before a Subcommittee on Appropriations House of Representatives*. 99th Cong., 1st Sess., Part 7. Washington, DC: Government Printing Office, 1985.

U.S. House of Representatives. *Defense Department Authorization and Oversight Hearings on H.R. 2287 [H.R. 2969] Department of Defense Appropriations for Fiscal Year 1984 and Oversight of Previously Authorized Programs before the Committee on Armed Services House of Representatives*. 98th Cong, 1st Sess., Part 4, Seapower and Strategic and Critical Materials Subcommittee Title I. Washington, DC: Government Printing Office, 1983.

U.S. House of Representatives. *Hearings before and Special Reports Made by Committee on Armed Services of the House of Representatives on Subjects Affecting the Naval and Military Establishments, 1965*. 89th Cong., 1st Sess. Washington, DC: Government Printing Office, 1965.

U.S. House of Representatives. *Hearings before the Subcommittee of the Committee on Appropriations House of Representatives*. 86th Cong., 2nd Sess., Department of the Navy. Washington, DC: Government Printing Office, 1958.

U.S. House of Representatives. *Hearings on Military Posture and H.R. 3818 and H.R. 8687 . . . before the Committee on Armed Services House of Representatives*. 92nd Cong., 1st Sess., Part 2. Washington, DC: Government Printing Office, 1971.

U.S. Naval Institute. *Naval Ordnance: A Textbook Prepared for Use of the Midshipmen of the United States Naval Academy*. Annapolis, MD: U.S. Naval Institute, 1933.

U.S. Navy. "Aegis Combat System." *Navysite*, accessed October 28, 2022. https://www.navysite.de/weapons/aegis.htm.

U.S. Navy. *A Report on Policies and Practices of the U.S. Navy for Naming the Vessels of the Navy*. Washington DC: Department of the Navy, 2013.

U.S. Navy. *Department of the Navy RDT&E Guide*. Washington, DC: Government Printing Office 1979.

U.S. Navy. *Register of Commissioned and Warrant Officers of the United States Navy and Marine Corps: July 1, 1939*. Washington, DC: Government Printing Office, 1939.

U.S. Navy. *Register of Commissioned and Warrant Officers of the United States Navy and Marine Corps: July 1, 1942*. Washington, DC: Government Printing Office, 1942.

U.S. Navy. *Register of Commissioned and Warrant Officers of the United States Navy and Marine Corps: July 1, 1944*. Washington, DC: Government Printing Office, 1944.

U.S. Navy. *Register of Commissioned and Warrant Officers of the United States Navy and Marine Corps: July 1, 1945*. Washington, DC: Government Printing Office, 1945.

U.S. Navy. "USS *Leahy* (DLG-16/CG-16)" (photograph). *Naval History and Heritage Command*, accessed September 30, 2022. https://www.history.navy.mil/content/history/nhhc/our-collections/photography/us-navy-ships/alphabetical-listing/l/uss-leahy—dlg-16-cg-16-0.html.

U.S. Navy, Bureau of Naval Personnel. *Duty Afloat for Engineering Specialists*. NAVPERS 10859-A. Washington, DC: Government Printing Office, 1960.

U.S. Navy, Bureau of Naval Personnel. "The General Line School Will Offer Further Professional Education." *Bureau of Personnel Training Bulletin*, March 15, 1946, 2–3.

U.S. Navy, Bureau of Naval Personnel. *Register of the Commission and Warrant Officers of the Navy of the United States, Including Officers of the Marine Corps*. Washington, DC: Secretary of the Navy, 1961. https://www.ibiblio.org/hyperwar/AMH/USN/Naval_Registers/1961.pdf.

U.S. Navy, Bureau of Naval Personnel, Training Division. *The Navy College Training Program V-12: Curricula Schedules/Course Descriptions*. 1943. https://www.google.com/books/edition/The_Navy_College_Training_Program_V_12/AiIDwgEACAAJ?hl=en&gbpv=1&printsec=frontcover.

U.S. Navy, Bureau of Ordnance. *Explosive Ordnance*. Washington, DC: Bureau of Ordnance, 1947. https://maritime.org/doc/ordnance/.

U.S. Navy, Naval Sea Systems Command. *Contracting Opportunities*, 2nd ed. Washington, DC: Naval Sea Systems Command Liaison Branch, 1983.

U.S. Navy, Naval Sea Systems Command. "Warfare Centers NSWC Corona Division." *NAVSEA*, accessed November 15, 2018.http://www.navsea.navy.mil/Home/Warfare-Centers/NSWC-Corona/Who-We-Are/Command-History/.

U.S. Navy, Naval Ship Systems Command. *Catalogue of Electronic Equipment, NavShips 900,116*, Suppl. 5, December 1952.

U.S. Navy, Office of the Chief of Naval Operations. *Catalogue of Advanced Base Functional Components*, 3rd ed. Washington, DC: Office of the Chief Naval Officer, 1945.

U.S. Senate. *Department of Defense Appropriations for Fiscal Year 1973, Hearings before a Subcommittee on Appropriations United States Senate . . . Part 3*. 96th Cong., 1st Sess. Washington, DC: Government Printing Office, 1979.

U.S. Senate. *Department of Defense Authorization for Appropriations for Fiscal Year 1980: Hearings before the Committee on Armed Services United States Senate*. 96th Cong., 1st Sess., on S.48. Washington, DC: Government Printing Office, 1980.

U.S. Senate. *Department of Defense Authorization for Appropriations for Fiscal Year 1981: Hearings before the Committee on Armed Services United States Senate*. 96th Congress, 2nd Sess., on S.2294, Part 6, Research and Development, Civil Defense. Washington, DC: Government Printing Office, 1980.

U.S. Senate. *Fiscal Year 1975 Authorization for Military Procurement, Research and Development, and Active Duty, Selected Reserve, and Civilian Personnel Strengths: Hearings before the Committee on Armed Services, United States Senate*. 93rd Congress, 2nd Sess., on S.3000. Washington, DC: Government Printing Office, 1974.

U.S. Senate. *Hearings before a Committee of the Committee on Appropriations United States Senate Ninety-Third Congress, First Session: An Act Making Appropriations for the Department of Defense for the Fiscal Year Ending June 30, 1974, and for Other Purposes*. Part 3, Department of the Navy. Washington, DC: Government Printing Office, 1973.

Wacker, Jerry E. "A Managerial Approach to the Determination and Selection of the Tactical Combat System for Surface Navy Ships," DTIC ADA039674. Defense Systems Management School, Port Belvoir, VA, May 1975. https://archive.org/details/DTIC_ADA039674/page/n3/mode/2up.

ORAL HISTORIES

Foster, Johns S., Jr. "Oral History Interview I," December 3, 1968, by Dorothy Pierce. LBJ Presidential Library, Austin, TX.

Hayward, John T. "History of the U.S. Naval War College: Reminiscences of NWC Presidency, 1966–68 and U.S. Navy Career." Interview by Dr. Evelyn M. Cherpak, May 1977, Naval War College.

Meyer, Wayne E. *The Reminiscences of Wayne E. Meyer, Rear Adm., USN (Ret.)* [oral history transcript of 18 interviews conducted by Paul Stillwell]. Annapolis, MD: U.S. Naval Institute, 2012.

Reich, Eli T. *The Reminiscences of Eli T. Reich, Vice Adm., USN (Ret.)*, 2 vols. [oral history transcript of interviews by John T. Mason Jr. (vol. 1, based on 11 interviews; vol. 2, based on 15 interviews)]. Annapolis, MD: U.S. Naval Institute, 1978.

INDEX

ABOUT THE AUTHOR

THOMAS WILDENBERG is an award-winning scholar with special interests in aviators, naval aviation, and technological innovation in the military. He is the author of a number of books on a variety of naval topics as well as biographies of Joseph Mason Reeves, Billy Mitchell, and Charles Stark Draper.

THE NAVAL INSTITUTE PRESS is the book-publishing arm of the U.S. Naval Institute, a private, nonprofit, membership society for sea service professionals and others who share an interest in naval and maritime affairs. Established in 1873 at the U.S. Naval Academy in Annapolis, Maryland, where its offices remain today, the Naval Institute has members worldwide.

Members of the Naval Institute support the education programs of the society and receive the influential monthly magazine *Proceedings* or the colorful bimonthly magazine *Naval History* and discounts on fine nautical prints and on ship and aircraft photos. They also have access to the transcripts of the Institute's Oral History Program and get discounted admission to any of the Institute-sponsored seminars offered around the country.

The Naval Institute's book-publishing program, begun in 1898 with basic guides to naval practices, has broadened its scope to include books of more general interest. Now the Naval Institute Press publishes about seventy titles each year, ranging from how-to books on boating and navigation to battle histories, biographies, ship and aircraft guides, and novels. Institute members receive significant discounts on the Press' more than eight hundred books in print.

Full-time students are eligible for special half-price membership rates. Life memberships are also available.

For more information about Naval Institute Press books that are currently available, visit www.usni.org/press/books. To learn about joining the U.S. Naval Institute, please write to:

Member Services
U.S. NAVAL INSTITUTE
291 Wood Road
Annapolis, MD 21402-5034

Telephone: (800) 233-8764
Fax: (410) 571-1703
Web address: www.usni.org